Advances in Anatomy
Embryology and Cell Biology

Vol. 152

Springer-Verlag Berlin Heidelberg GmbH

F. Eckstein   B. Merz   C. R. Jacobs

Effects of Joint Incongruity on Articular Pressure Distribution and Subchondral Bone Remodeling

With 53 Figures and 2 Tables

 Springer

FELIX ECKSTEIN, PD Dr. med.
Ludwig Maximilian University Munich
Musculoskeletal Research Group
Institute of Anatomy
Pettenkoferstrasse 11
80336 Munich, Germany
(e-mail: eckstein@anat. med. uni-muenchen. de)

BEAT MERZ, Dr. sc.
Institut Straumann
4437 Waldenburg, Switzerland

CHRISTOPHER R. JACOBS, Associate Professor
Musculoskeletal Research Laboratory
Department of Orthopedics
Pennsylvania State University
Hershey Medical Center
Hershey, PO Box 850, PA 17033, USA

ISSN 0301-5556

Library of Congress-Cataloging-in-Publication-Data
Eckstein, Felix H.: Effects of joint incongruity on articular pressure distri-
bution and subchondral bone remodelling; with 2 tables / F. H. Eckstein;
B. Merz; C. R. Jacobs. – Berlin; Heidelberg; New York; Barcelona; Honkong;
London; Milan; Paris; Singapore; Tokyo: Springer, 2000
    (Advances in anatomy, embryology, and cell biology, Vol. 152)
    ISBN 978-3-540-66212-9   ISBN 978-3-642-57184-8 (eBook)
    DOI 10.1007/978-3-642-57184-8

© Springer-Verlag Berlin Heidelberg 2000
Originally published by Springer-Verlag Berlin Heidelberg New York in 2000

Printed on acid-free paper – SPIN: 10698790   27/3136wg - 5 4 3 2 1 0

*Dedicated to My Wife Eva-Maria
and My Daughter Franziska*

Contents

1 Introduction and Questions

The synovial joints play a central role in the positioning of the individual relative to the environment and in locomotion. They also enable the more advanced species to make active changes in the environment and, by means of gesticulation and "body speech," to exchange information. The integrity of the joint as an organ is thus an essential prerequisite for the efficient functioning, survival and well-being of the organism. The purpose of the joints is to allow the relative movement of the limb segments in certain directions and to limit it in others, in order to provide sufficient stability to the limbs. In addition to their kinematic function, they also transmit the forces occurring during static and dynamic activity, which result from the external loading and the musculature, from one segment of the body to another.

Since most muscles have a much shorter lever arm than the forces exerted from outside, the joint reaction force is in general several times higher than the body weight. Therefore the tissues which form the joints are subjected to considerable mechanical stresses and strains. However, the structure of the joints and the material properties of the connective tissues allow them, under ideal conditions, to maintain their mechanical functions adequately for decades in spite of the high loading. This is at least in part achieved by the growth, differentiation and regeneration processes being under the control of regulatory feedback mechanisms which guarantee their dynamic adaptation to the mechanical requirements, and an optimization of their functions during daily use.

Degenerative diseases of the joints have become tremendously more common during the twentieth century as a result of the considerable elongation of life expectancy. Radiographically, most persons over the age of 65 years suffer from osteoarthrosis in at least one joint (Felson 1988, 1990). Musculoskeletal diseases are responsible for more than twice as many days lost at work than the diseases of any other organ system (sickness statistics of the German health insurance company AOK for the year 1992). Degenerative disease of synovial joints produced costs of approx. $65 billion in the United States in 1992 (Yelin and Callghan 1995) and the costs are expected to reach 1% of the gross national product at the end of the century (Yelin 1998). Osteoarthrosis is the cause of over 50% of chronic disease among those over 65 years of age (Lidgren 1996), second only to cardiovascular disease in causing severe chronic disability (Badley et al. 1978; Peyron 1986). Epidemiological investigations show that mechanical factors play a dominant role in the initiation and progression of osteoarthrosis (Peyron 1986; Felson 1988; Radin et al. 1991; Mow and Ratcliffe 1997). In spite of the enormous social and economic significance of degenerative joint conditions (Yelin and Callghan 1995; Yelin 1998), the disease has so far remained unsatisfactorily defined, and the mechanical and cellular processes which are responsible for the tissue damage are not precisely understood (Ostergaard and Salter 1998).

Animal experiments have shown that mechanical stress which exceeds a particular threshold causes local cell and tissue changes in the articular cartilage similar to those observed in human osteoarthrosis (Simon et al. 1972; Hoch et al. 1983; Radin et al. 1978, 1991; Zimmermann et al. 1988; Saxena et al. 1991; Tomatsu et al. 1992). Since local stress peaks quite obviously have a harmful effect on the tissues, it appears necessary for the mechanical forces to be transmitted relatively homogeneously throughout the joint, and the pressure spread uniformly across the articular surface. In this connection it has often been taken for granted that the joint components make a perfect fit. On the other hand, however, numerous qualitative observations show a regular deviation from such an ideal fit in humans (e.g., Walmsley 1928; Goodfellow and Bullough 1967; Bullough et al. 1968, 1973; Afoke et al. 1980, 1984a; Bünck 1990; Iannotti et al. 1992).

In this context, it is necessary to define how the absolute shape of the joint components or their relative fit (i.e., their differences from one another) should be described. The concept of asphericity, for instance, refers to the absolute shape of an articular surface and to its departure from a true sphere. Here no statement about the fit of one joint surface to the other is possible. The relative shapes of two joint components are generally described as congruous or incongruous. Congruity refers to an ideal fit, with all regions of the joint surfaces being in contact with each other. Incongruity, on the other hand, indicates a relative difference in the shapes, so that – at least when the joint is unloaded – only parts of the surfaces come together, while in other places there remains a space between them.

Various terms have been used for describing the absolute and relative form of the joint components. The concept of joint incongruity in clinical – predominantly in the radiological and orthopedic – terminology usually refers to the bone-cartilage interface of the articulating surfaces in a radiograph, or, more precisely, to the transition between the calcified and uncalcified cartilage. Departures from the regular form of the components are described which are usually judged to be pathological, and which lead to overloading of the cartilage and to degenerative, osteoarthrotic changes. "Radiological incongruity" is therefore frequently equated with the term "prearthrotic deformity" (Hackenbroch 1923).

At this point it is necessary to define "physiological joint incongruity." This refers to the subtle departures from an ideal fit found under natural conditions which are of no direct pathological significance. They cannot be observed by radiography or computed tomography (CT) because these techniques do not directly delineate the cartilage. Since the cartilage is not of equal thickness in all regions, and there may be substantial inhomogeneity in the distribution (Werner 1897; Kurrat and Oberländer 1978; Müller-Gerbl et al. 1987; Ateshian et al. 1991, 1995; Soslowsky et al. 1992a; Eckstein et al. 1992; Milz et al. 1995, 1997; Adam et al. 1998), the form of the joint surface can differ markedly from that of the bone-cartilage interface. In the present monograph, the concepts and observations invariably refer to the form of the cartilaginous articular surface and not to that of the bone-cartilage interface.

Pauwels (1963) postulated that the distribution of the subchondral mineralization observed in lateral radiography of the trochlear notch represents a "materialized stress diagram." He assumed that the humeroulnar joint is congruous, and compared the density pattern with the pressure distribution at the joint surface at various degrees of flexion. This hypothesis is interesting since it approaches the idea that the long-term stress distribution acting on the joint surface, which is potentially responsible for the degenerative processes, can be derived from the quantitative distribution of the bone tissue by means of noninvasive radiological methods.

Later investigations, however, have shown that the distribution pattern of subchondral bone density – particularly that of the trochlear notch, but also that of the hip and ankle joints – accords better with the assumption of joint incongruity than with that of precise congruity, and that hence the incongruity implies important functional consequences on normal tissue adaptation (e.g., Tillmann 1971, 1978; Oberländer 1973; Müller-Gerbl et al. 1993; Müller-Gerbl and Putz 1995; Müller-Gerbl 1998). Based on the idea that the subchondral mineralization can be regarded as a morphological reflection of the pressure acting on the joint surface, it has been concluded that the incongruity affects the long-term stress distribution over the articular surface, and also that age-dependent changes in the joint incongruity (Bullough et al. 1968; Riede et al. 1971) can be derived from subchondral mineralization patterns. Since it is now possible to determine the distribution of subchondral bone density noninvasively by computed tomography (CT) osteoabsorptiometry (OAM) (Müller-Gerbl et al. 1989, 1992; Müller-Gerbl 1998), this method may potentially permit conclusions about the long-term distribution of forces in the joint and, finally, about the mechanical etiology of osteoarthrosis.

These theories are based on the assumption that the distribution of stress at the joint surface and in the articular cartilage corresponds very closely to its distribution in the subchondral bone. However, Simkin et al. (1980) have suggested that tensional stress can appear in the subchondral bone of the concave joint component, although they did not take into account joint incongruity. In this context it is relevant that in incongruous joints where the pressure is inhomogeneously distributed, bending moments can appear in the concave joint component, and that the stress and strains of the subchondral bone may therefore differ significantly from the contact pressure at the articular surface. Simkin et al. (1980, 1991) based their views on the fact that the subchondral plate of the concave joint component is thicker than that of the corresponding convex component, but did not offer precise quantitative estimates of the assumed tensional stress.

The question arises as to what extent the natural incongruity of synovial joints affects the pressure transmission through the joint, and what significance these mechanical relationships have for the mechanoadaptive and degenerative processes taking place in the cartilage and in the subchondral bone tissue. An experimental examination of the interaction between joint incongruity, pressure transmission, and functional adaptation of the bone is for several reasons problematic. Whereas the ions and molecules that take part in the regulation of cellular processes can today be accurately analyzed with biochemical, cell biological, and molecular biological techniques, mechanical signals (i.e., strains and stresses in the material) cannot be measured in biological tissues without introducing artifacts. If, for instance, the pressure distribution over a joint surface is measured with a pressure-sensitive film, it must be accepted that the film to some extent distorts the actual fit of the joint (the congruity or incongruity), and that under certain circumstances the value of the information is seriously limited. It is even more difficult, and as yet technically impossible, to obtain an artifact-free measurement of the mechanical strain in the subchondral bone. Furthermore, the fit of a joint can be experimentally altered only in an imprecise manner and very approximately, and the relevance of experiments on the physiological incongruity is therefore questionable. A third point is that the biological processes of tissue adaptation require long periods of time, making the comparative (parametric) and systematic examination of various influential components intrinsically difficult.

Mathematical models and computer simulation offer a modern alternative to the experiment. The rapid increase in computer power during the past two decades has made computational analyses increasingly attractive, and this has made it possible to estimate the mechanical strains and stresses in the tissue and, even more importantly, to simulate the process of tissue adaptation (Huiskes and Chao 1983; Huiskes and Hollister 1993; Mow et al. 1993; Prendergast et al. 1997a). Computer simulations have the advantage of allowing mechanical signals to be calculated in situations in which artifact-free measurements are not possible, and they can also predict processes of mechanobiological tissue adaptation in a time-effective manner. Single factors can be isolated in a complex nonlinear system, and their relative effect on the system as a whole be analyzed. It must, however, be remembered that the predictive reliability of these models depends largely on basic assumptions about the quantitative distribution of the tissue, its material properties, and the rules of adaptation at a cellular level. Important factors for the construction of a model therefore include an accurate consideration of the anatomical relationships (e.g., the extent of the incongruity), the use of algorithms of which the validity or predictive capacity has been ascertained (for instance, in animal experiments) and confirmation of the plausibility of the prediction in terms of established morphological or experimental data. Once these facts are taken into account, computer simulation is an extremely efficient method of gaining insight into the interdependence of morphological structure, physiological function, and mechanobiological regulatory mechanisms which cannot be presently achieved in the same way with other techniques.

For analyzing the effect of physiological incongruity on pressure transmission and the adaptation in human joints, in the present study we select the method of computer simulation. Particular importance is attached to validating the predictions of the model in terms of morphological and experimental evidence.

The questions asked can be grouped under the following thematic complexes:

Incongruity and Pressure Distribution. How does the physiological incongruity affect the pressure at the joint surface? What is the relationship between the joint incongruity, the magnitude of loading, and the stress distribution over the articular surface?

Incongruity and Subchondral Bone Remodeling. Does simulation of bone remodeling result in a change in the pattern of subchondral bone density in models with different types and degrees of incongruity? Do joints with different types of fit show differences in their radiographic subchondral mineralization pattern? Do the predictions of the computer simulation agree with the distribution pattern determined by experiments?

Pressure Distribution and Subchondral Bone Remodeling. What is the quantitative relationship between the pressure distribution at the articular surface, the pressure distribution in the subchondral bone, and the subchondral bone density/mineralization?

The Value of CT Osteoabsorptiometry. Can the incongruity of a joint and/or its long-term mechanical loading (average long-term magnitude and direction of the joint reaction force) be predicted noninvasively from the pattern of subchondral mineralization by means of CT-OAM (Müller-Gerbl et al. 1989, 1992; Müller-Gerbl 1998)?

Incongruity and Subchondral Tension. Does tensional stress (but not compression alone) play a role in the functional adaptation of the subchondral bone? If so, what is the magnitude of the tensional stress in relationship to the compressive stress?

Morphological Correlates of Subchondral Tension. Is their a functional consequence of the appearance of tensional stress in the subchondral bone? Is the tensional stress large enough to cause a specific adaptation of the subarticular trabecular orientation and the course of the collagen fibrils in the subchondral plate?

2 Basis for the Design of the Computer Models

2.1
Qualitative and Quantitative Description of Joint Incongruity

As noted above, various terms are used to describe the form of joint surfaces. It must be decided whether the geometric form of a single joint surface (e.g., its departure from the section of a sphere: asphericity) or the relative shapes of two articulating surfaces (congruity/incongruity) is to be described. The prerequisite for the qualitative or quantitative determination of joint incongruity is that the articulating surfaces must be brought into real or virtual contact. This must take place with minimal force, since otherwise the cartilage is deformed, and the original fit of the articulating surfaces, which is necessary for constructing a model cannot be determined. If the joint components are brought into contact with minimal force in a certain position, and if the contact area is identical with the entire surface of the smaller partner, the joint can be regarded as congruous in this position. If only parts of the surfaces make contact, with a joint space remaining elsewhere, the joint is incongruous in this position. (It should be repeated that the term "joint space" here refers to the distance remaining between the cartilaginous surfaces, and not – as in clinical terminology – the bone-cartilage interfaces.) The magnitude and distribution of the joint space width (under the minimum force necessary for joint contact) can be regarded as a quantitative measure of the incongruity in that position of the joint.

Various types of joint incongruity can be distinguished depending upon the distribution pattern of the joint space width. If, for instance, the joint partners are centrally in contact, with the joint space width increasing towards the periphery, this geometric configuration can be regarded as "convex incongruity" (Eckstein et al. 1995a). If, however, there is contact peripherally, with the space increasing towards the center, we can speak of "concave incongruity" (Eckstein et al. 1995a).

Historically, the descriptions of joint incongruity have been based on very different techniques. In the following sections the essential findings and accepted methods of important studies on joint incongruity will be presented.

2.1.1
Hip Joint

Walmsley (1928) and MacConnail (1950) reported that the articular surfaces of the human hip joint do not make an ideal fit, and that these must be regarded as incongruous. In a classical investigation, Bullough et al. (1968) measured the radius of the acetabulum and femoral head in 53 hip joints with mechanical callipers. Based on each of four measuring points, the difference between the maximal and minimal radii of the surfaces (divided by the mean radius) was used as a quantitative measure of asphericity. It was shown that neither of the articulating surfaces represented a section of a true sphere, that the acetabulum had on average a greater degree of asphericity than the femoral head, and that the asphericity of each surface became less as age increased. In simplest terms, Bullough et al. (1968) compared the femoral head with a ball lying in a Gothic arch – the acetabulum. Although this gives some idea of the incongruity of the hip joint, it must be pointed out that it does not relate to the actual incongruity of individual joints in a particular position, since the fit between each component and its partner was not examined.

Later investigations revealed that, in those positions of the hip joint which occur during normal walking, the contact areas take up only a part of the articular surface under light loading, while significant areas of the lunate surface make no contact at all (Greenwald and O'Connor 1971; Greenwald and Haynes 1972; Bullough et al. 1973; Goodfellow and Mitsou 1977; Afoke et al. 1980). These results indirectly suggest that there is a permanent incongruity of the joint components for the respective positions of the joint. They have been extended by the observation of an inhomogeneous pressure distribution in the hip using pressure sensors or pressure-sensitive films (Day et al. 1975; Mizrahi et al. 1981; Brown and Shaw 1983; Miyanaga et al. 1984; Adams and Swanson 1985; Afoke et al. 1987). However, it must be remembered that some distortion of the actual incongruity by the pressure-sensitive devices cannot be excluded. It should also be noted that very different descriptions of the position and size of the contact and load-bearing areas are given in the studies cited above. It has been reported that their location varies not only between individuals, but also within the same individual, depending on the position of the joint and the magnitude and direction of the joint reaction force. Other studies have investigated the shape and pressure transmission in the hip joint by special endoprostheses equipped with ultrasonic transducers and pressure sensors (Rushfeldt and Mann 1979; Rushfeldt et al. 1981; Hodge et al. 1989). However, it was the asphericity and not the incongruity of the acetabulum that was measured since the relative distortion of the acetabulum was determined relative to an artificially constructed spherical femoral head, and not to the normal joint component.

2.1.2
Ankle Joint

Regular differences in shape of the joint components have also been described in the ankle joint (Riede et al. 1971; Wynarsky and Greenwald 1983), with the trochlea tali showing a greater depth than required for a precise fit with the distal tibia. It has been shown that the "talus profile ratio," which is a measure of the depth of the socket, gets less with increasing age, and in elderly individuals complete congruity can be observed (Riede et al. 1971).

2.1.3
Humeroulnar Joint

Using a staining method, Goodfellow and Bullough (1967) described peripheral (ventrodorsal) contact areas in a specimen of the humeroulnar joint of a young man. These were located in the parts of the joint surface adjacent to the coronoid process and olecranon, but no contact was found in the depth of the trochlear notch. A similar ventrodorsal pattern of contact was found in investigations using anatomical sectioning (Bullough and Jagannath 1983) and this was confirmed by the direct comparison of different experimental methods (Stormont et al. 1985).

A condition common to these three joints is that, in each case, the concave component is deeper than required to fit perfectly with the convex component. They can therefore be referred to as "concavely" incongruous.

2.1.4
Humeroradial Joint

In the humeroradial joint it has been shown using a casting method that the capitulum articulates with the radial fovea at its center (Bünck 1990). The measurement of anatomical sections revealed that the fovea has a greater radius of curvature than the corresponding capitulum (Bünck 1990). This joint can therefore be said to show "convex" incongruity.

2.1.5
Shoulder Joint

Conflicting results are available for the fit of the articular surfaces of the shoulder joint. Pauwels (1959, 1965, 1980) concluded indirectly from an analysis of the split lines of the surface layer of the glenoid cartilage that the radius of the humeral head must be greater than that of the socket. Soslowsky et al. (1992a) derived average radii of curvature for the humeral head and glenoid cavity by a stereophotogrammetric method; they reported both surfaces to be nearly spherical and regarded the glenohumeral joint as virtually congruous. In a parallel investigation (Soslowsky et al. 1992b) without averaging of the geometric data but exact reconstruction of the individual joint surfaces, it was reported, however, that the calculated glenohumeral contact areas cover only a small section of the cavity and are subjected to some displacement during abduction. Both Ianotti et al. (1992), who based their findings on anatomical sections and magnetic resonance (MR) images, and Kirsch et al. (1993), who used a casting technique, reported that the glenoid cavity has a greater radius of curvature than the humeral head, and that the humerus is therefore capable of a certain amount of translation within the socket.

2.1.6
Our Own Investigations

The precise analysis of the significance of joint incongruity for pressure transmission and the adaptive processes in the subchondral bone requires a quantitative evaluation of the joint incongruity which can then be included in the design of a corresponding model. In our experiments the relative deviation of the joint components in a selected position was regarded as a measure of its incongruity, using the minimal loading necessary to bring the components into contact. Quantification of the incongruity was achieved by determining the surface distribution of the width of the joint space with a casting method, the thickness of the cast being always measured perpendicular to the articular surface of the convex component.

First, a polyether cast was used to determine the quantitative distribution of the width of the joint space (Eckstein et al. 1993) and the contact areas (Eckstein et al. 1994b) in the humeroulnar joint of an elbow at a 90° flexion angle. It was found that the trochlear notch (where the diameter of the socket is about 20 mm) is generally some 0.5–2.0 mm deeper than would correspond to an exact fit with the humeral trochlea. This is equivalent to a departure of the shape of the notch from a true cylinder of about 5%–20%. The incongruity increases from the central sagittal ridge towards the periphery of the articular surface (Fig. 1) and shows a certain dependence upon the type of the articular surface. In the majority of cases the cartilage surface of the notch is interrupted in its depth (Tillmann 1971, 1978; Fig. 2), and it is only occasionally that the joint surface is continuous. In this latter case the incongruity of the joint is less than in specimens with a subdivided joint surface, and rarely it may actually be congruous. Tillmann (1971, 1978) attributed this division of the articular surface to the incongruity of the humeroulnar joint.

A subsequent study (Eckstein et al. 1995a) examined humeroulnar joints with divided articular surfaces at 30°, 60°, 90°, and 120° of flexion. The quantitative descriptions of the deeper notch and the ventrodorsal contact areas were essentially confirmed for all degrees of flexion. This indicates that the humeral surface is almost circular in a sagittal section, and that the degree of humeroulnar incongruity is relatively independent of the specific joint position.

The findings obtained in these biomechanical investigations have recently been qualitatively confirmed in intact joints by high-resolution MRI (Eckstein et al. 1996a). In the humeroulnar joint these demonstrate the deeper socket (Fig. 3a, concave incongruity) and in rare cases a congruous relationship between the joint components (Fig. 3b). Depending upon the subject selected and the angle of flexion in the humeroradial joint there are cases of congruity of the components (Fig. 3c) and others in which the fovea has a larger radius of curvature than the corresponding capitulum (Fig. 3d, convex incongruity).

Quantitative data on joint incongruity have also been obtained in the hip by simulating the typical loading during the stance phase (Eckstein et al. 1997a) and during four phases of the gait cycle (heel strike, midstance, heel off, and toe off; von Eisenhart-Rothe et al. 1998, 1999). It has been demonstrated that under small loads it is primarily the periphery of the lunate surface that makes contact, whereas a relevant joint space remains at the center of the articular surface (concave incongruity). This incongruity has also been shown to lead to an inhomogeneous distribution of pressure, with maxima at the periphery and in both the anterior and posterior horns of

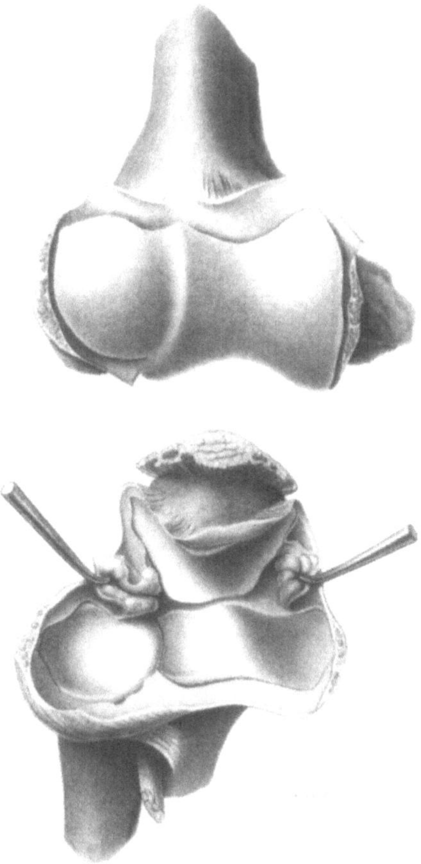

Fig. 1. Ventral view of the human elbow joint. The trochlear notch shows a main sagittal ridge, its articular surface being transversely divided in the depth of the notch (drawn by Cornelia Pankalla). With permission from Eckstein F, Löhe F, Hillebrand S, Bergmann M, Schulte E, Milz S, Putz R (1995) Morphomechanics of the humero-ulnar joint: I. Joint space width and contact areas as a function of load and flexion angle. Anat Rec 243:31896326 (Copyright Wiley-Liss, Inc., a division of John Wiley & Sons Inc., 1995)

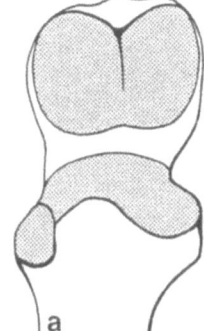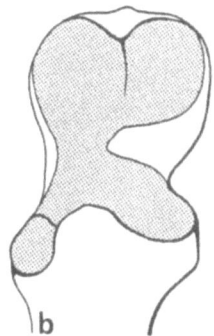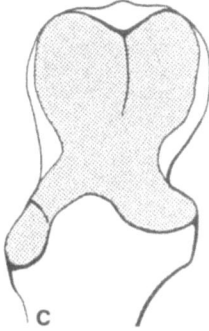

Fig. 2a–c. Morphology of the articular surface of the trochlear notch according to Tillmann (1971, 1978). **a** Completely divided joint surface (found in approx. 65% of all adults). **b** Medially divided joint surface (found in approx. 30% of all adults). **c** Continuous joint surface (found in approx. 5% of all adults). With permission from Eckstein F, Lohe F, Schulte E, Muller-Gerbl M, Milz S, Putz R (1993) Physiological incongruity of the humero-ulnar joint: a functional principle of optimized stress distribution acting upon articulating surfaces? Anat Embryol 188:449–455 (Copyright Springer 1993)

11

Fig. 3a–d. Assessment of the physiological incongruity in the humeroulnar and the humeroradial joint with MRI (T1-weighted, high-resolution fat-suppressed gradient echo sequence; TR=45 ms, TE=12 ms, FA=40°; resolution $1\times0.2\times0.2$ mm^3). **a** "Concave" incongruity of the humeroulnar joint. **b** Rare case of congruity in the humeroulnar joint. **c** Congruity of the humeroradial joint. **d** "Convex" incongruity of the humeroradial joint. (With permission from Eckstein F, Merz B, Sittek H, Kolem H, Reiser M, Putz R (1996a) Geometry-to-pressure relationship in the human elbow joint – a qualitative analysis using MRI and finites elements. Eur J Anat 1:23–30 (Copyright Soc. Anat. Española 1996)

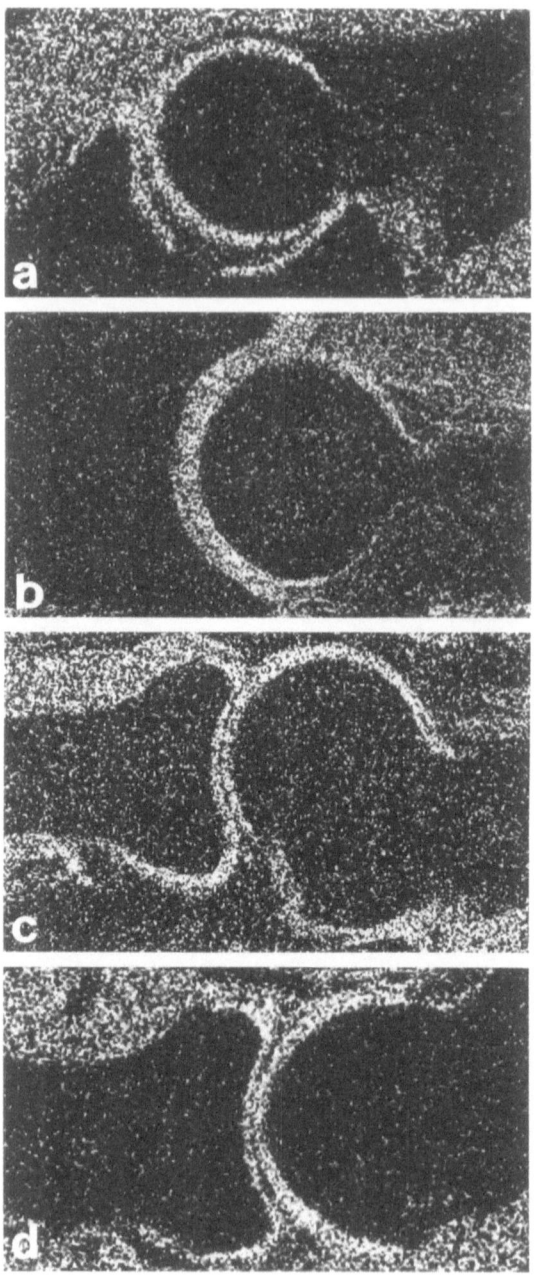

the lunate surface, rather than in the acetabular roof (von Eisenhart-Rothe et al. 1997, 1998, 1999).

In the shoulder, we have demonstrated with a casting technique that in about 50% of joints the contact areas are located exclusively in the center of the glenoid during abduction (convex incongruity), and that contact throughout the entire joint surface (congruity) or at the superior and inferior aspects (concave incongruity) is established in the others (Eckstein et al., unpublished data). This corresponds with either central or superoinferior pressure maxima in the glenoid, recorded with Fuji film (Conzen et al. 1997; Conzen and Eckstein 1999). At 120° of flexion of the shoulder the contact areas and pressure distribution were found to be consistent with that during abduction, but upon 90° external rotation (during abduction) the contact and load-bearing areas always occupied a more central location (convex incongruity) (Conzen et al. 1997; Conzen and Eckstein 1999; Eckstein et al. unpublished data).

The elbow joint presents an almost perfect model of a physiologically incongruous joint since concave incongruity is shown by its humeroulnar, and convex incongruity by its humeroradial component. Because of the relatively high number of previous investigations into the incongruity and the subchondral bone density (Pauwels 1963; Goodfellow and Bullough 1967; Tillmann 1971, 1978; Bullough and Jagannath 1983; Bünck 1990; Eckstein et al. 1993, 1995a, 1996a), on which basis accurate models can be designed, the current study focuses on the elbow joint. Since other joints also show similar geometric relationships, the results of this investigation can within certain limits be transferred to these.

2.2
Finite Element Method

Computer simulation methods offer decided advantages for investigating the relationship between joint incongruity, pressure transmission, and the adaptation of subchondral bone. The most frequently used numerical method employed is the finite element (FE) method. This was originally used in engineering design, particularly in the aircraft industry. Because of its distinct advantages, however, use of the FE method in biomechanical and orthopedic research has increased enormously over the past two decades (Huiskes and Chao 1983; Mow et al. 1993; Huiskes and Hollister 1993; Prendergast 1997a). As noted above, the FE method is not an experimental method and is not based – as is the strain gauge technique, for instance – on direct measurements. The objective of the FE method is to theoretically predict the mechanical behavior of a structure given its shape and material properties, and the loading to which it is exposed. The corresponding physical problem – the deformation in a bone or joint under load, for instance – is formulated as a series of differential equations with certain initial and boundary conditions. With the exception of special simple cases, it is not in general possible to arrive at an exact analytical solution of the problem which satisfies every point of the system. The FE method is used to calculate approximate solutions which satisfy the differential equations as a weighted average. In this way a complex problem is broken down into a collection of smaller and simpler problems which can be solved numerically. The three main phases of a static (time-independent) FE structure analysis are presented below:

2.2.1
Preprocessing

During preprocessing, the model (a bone, for instance, or a joint) is defined as a two- or three-dimensional object and segmented into individual elements which are bound together at nodal points. In the two-dimensional case the elements usually take the form of triangles or quadrilaterals, and in the three-dimensional case that of tetrahedra and hexahedra. The corners of the elements are defined by the associated nodes, with adjacent elements sharing the nodes at their points, lines or surfaces of contact.

After the creation of such a mesh in the area or volume of interest – the so-called discretization – the boundary conditions of the model are defined. Corresponding to the biomechanical load case, the possible displacements are set to zero for certain nodes (the model being fixed here), whereas at others, forces or displacements are applied. In addition to the geometry and boundary conditions, the material properties of the investigated structure must be defined. These values (namely the elastic modulus and the Poisson's ratio, see below) determine the relationship between the known external forces and the unknown displacement of the nodes in the model, and are assembled into the so-called stiffness matrix. The description of the material properties is either the same for all the elements (homogeneous) or differs between individual elements or groups (heterogeneous). The material properties must be obtained by experimental examination; approximations can, however, be derived from the attenuation values of the bone tissue in CT, for instance.

For the sake of simplicity, bone is often regarded as an isotropic, linear elastic material. This presumes that the tissue behaves mechanically as a continuum, and that the stress and strain are linearly related (Hook's law). It further assumes that the material shows the same stiffness at compression and tension, and that the stiffness is equal in all directions (isotropy). Such a material can be characterized by two elastic constants, these are typically the elastic modulus (or Young's modulus) and the Poisson's ratio. In an isotropic material both of these can be determined experimentally in a uniaxial tension or compression experiment. The elastic modulus corresponds to the slope of the linear region of the stress-strain curve. This is expressed in units of stress (pascals) and provides the theoretical stress value for extending a test specimen to twice its original length. Poisson's ratio is defined as the relationship of the strain in the direction of an applied load to the strain acting perpendicular to this direction. It therefore measures the tendency of the material to extend laterally in a compression test. The dimensionless Poisson ratio can take on any values between 0 (no lateral extension) and 0.5 (maximum lateral extension=incompressible material).

Experimental results have established that bone tissue shows almost identical elastic properties under tension and compression (Carter et al. 1980), but that the assumption of isotropy (independence of Young's modulus from the spatial direction) is not valid. Because of their microstructural organization cortical and trabecular bone are anisotropic; that is to say, that the material properties depend upon the direction of loading (Reilly and Burstein 1975; Brown and Ferguson 1980). Determining the anisotropic constants is technically difficult, however, and the anisotropy of the tissue is therefore frequently neglected. In any case, it is not permissible to assume that the material properties of bone are homogeneous. Various investigations have revealed the important influence of the density (or the CT values) on the local value of Young's modulus (Carter and Hayes 1977; Gibson 1985; Rice et al. 1988; Hodgskinson and Currey 1990; Lotz et al. 1990; Rho et al. 1995). Estimations of the percentage influence of density on stiffness vary from 72% (Hodgskinson and Currey 1990) to 90% (Lotz et al. 1990). It is therefore justifiable, depending on which question is formulated, to neglect the anisotropy of bone, provided the inhomogeneity of the tissue is taken into account (Hayes et al. 1982).

2.2.2
Processing

During the second phase of the FE analysis (processing) the actual displacements of all the nodes under load are calculated in the model. As noted above, the mathematical solution does not provide the exact conditions for equilibrium for each of the points but a weighted mean equilibrium for the whole structure with minimal local error residuals. Regarding numerical accuracy, the best results are achieved by using the largest number of elements or nodes possible. This is very costly, however, in terms of computer time and required data storage, and therefore a compromise must be achieved between the two requirements in each case. A finer mesh is required only in areas with steep stress/strain gradients; therefore the refinement of the mesh usually varies throughout the model. Whether the mesh is sufficiently fine can be checked by a convergence test. To do this, the mesh is repeatedly made finer, until the results in the region of interest are sufficiently similar for two different stages of this process. When this has been achieved, one can describe the solution as having reached convergence, since additional improvement in accuracy obtained by further refinement of the mesh would not be significant. Numerical accuracy is obviously only one of the many aspects of the quality of the model. Factors such as the validity of the biomechanical assumptions, adequate boundary conditions and material models can have a much greater influence.

2.2.3
Postprocessing

During the third phase (postprocessing) a graphic output is obtained of the displacement of the nodes or secondary values (e.g., the stresses and strains in the material). The strains are calculated as the difference between the current length of an arbitrary distance (under mechanical loading) and its original length (before mechanical loading), divided by the original length. The strains are expressed as percentages, a negative value corresponding to a compressive and a positive value to a tensional strain. The stress is a measure of the intensity of a force acting within a body. The normal stress in any arbitrarily selected (imaginary) surface in the material can be calculated from the force acting on this surface and the size of the surface. Stress is measured in the SI system of pascals (newtons per square meter; 1 MPa=1 N/mm^2). It is convenient to express the local stress and strain by dividing it into its components. This can be done in terms of an imaginary cube within the material, whereby the forces acting on its surfaces are divided into three normal components (perpendicular to the surfaces of the cube) and three shear components (parallel to the surfaces of the cube). These components (which together constitute a strain or stress tensor) can be graphically displayed in any spatial direction. It is more useful, however, to arrange the coordinate system of the cube at each point of the analyzed structure in such a way that each normal component assumes a maximal or minimal value and the shear components vanish. With this procedure one obtains the so-called principal stresses, for example, tension and compression, which can be displayed as vectors. This provides an impression of the flow of forces through the object.

Furthermore, there are scalar (directionless) values which describe the effective stress acting in the material. One can obtain the dilatational stress (the hydrostatic pressure), which is the component of stress that produces a change in volume of the material, from the sum of the normal stress components divided by the number of these components. In contrast, the "von Mises" stress provides a measure of the shear stress – or deviational stress – which is associated with changes in the form and not the volume of the material. Historically, the von Mises stress was developed in engineering applications to predict the begin of the plastic deformation and the failure of metals, since the yield stress of a metal is independent of dilatational stress, and dilatational stress does not contribute to von Mises stress. However, since both the dilatational stress associated with the changes in volume and the shear stress associated with the changes in form contribute to the failure and to the functional adaptation of bone, neither of these values alone is suitable for predicting the remodeling processes of the bone tissue (see above). A scalar quantity, however, which includes both the dilatational and the deviational stress is the strain energy density (SED). This provides the deformation energy per unit volume of the material, and corresponds to the area under the stress-strain curve at a given strain. It can be calculated by multiplying the stress tensor (i.e., the matrix of the normal and shear stresses) and the strain tensor (i.e., the matrix of the normal and shear strains divided by two). Many current bone-remodeling algorithms use the SED as a feedback value (or mechanical signal) to predict the functional adaptation of the bone tissue.

2.3
Theory of Mechanoadaptive Bone Remodeling

2.3.1
Roots, Mathematical Formulation, and Implementation with the Finite Element Method

As early as 1638 Galileo (cited after Ascenzi 1993) investigated the causal formation of bone, and noted in his work *Discorsi e dimostrazioni mathematiche, intorno a due nuove scienze* that the bones must be disproportionately wide in larger animals to ensure sufficient support for their weight. Bourgery (1832) was one of the first to describe the fine structure of the trabecular bone, and theorized that bone exhibits the greatest possible degree of strength and lightness. This so-called "maximum-minimum law," whereby maximal strength of the structure is achieved while using the minimal amount of material, was similarly formulated by Bell (illustration in *Paley's Natural Theology*, 1834). The anatomist von Meyer (1867) observed the astonishing similarity between the trabecular architecture of the proximal femur to the stress trajectories in a bent girder (known in his time as a Fairbairn crane). The stress trajectories had been determined by the engineer Culmann on the basis of graphic

16

statics, a method that had been developed at that time. This was described by Roesler (1987) as the first cooperative multidisciplinary effort in biomechanics. Wolff (1892) put forward the "trajectorial theory" of the trabecular bone, according to which the trabeculae follow the course of the maximal compressive or tensional (the principal) stresses. Following Culmann's analysis, he claimed that the trabeculae must always cross each other at a right angle, and on this basis corrected von Meyer's drawing of the trabecular architecture of the proximal femur. Although his description was essentially qualitative, he expressed the opinion that there is a definite mathematical relationship between the structure of bone and its mechanical stress. Roux (1881) adopted the idea from Lamarck (1809) and Darwin (1872) that the structure of an organ represents an adaptation to its normal function. He claimed that the bone structure adapts itself to the latest local mechanical conditions by means of a cell-controlled regulating mechanism (theory of functional adaptation), and (at first in disagreement with Wolff's belief) that this process also takes place in the absence of pathological changes and in the mature adult. According to Roesler (1987), these concepts (the maximum-minimum law, the trajectorial hypothesis, and the theory of functional adaptation) finally welded together in the "law of bone transformation" (Wolff 1892) or, more shortly, "Wolff's law."

Many attempts have since been made to describe these rules in a quantitative mathematical form (Koch 1917; Thompson 1942; Frost 1964). Pauwels (1955, 1965, 1980) and Kummer (1962, 1963, 1972) compared bone with a feedback system that reacts to mechanical underloading or overloading with bone resorption (atrophy) or bone deposition (hypertrophy), respectively. It was assumed that a regular standard or reference value of the stress in the bone is striven for, at which no net deposition or resorption takes place, or where these processes are in equilibrium. According to this theory the degree of remodeling would be determined essentially by the difference between the actual stress present and an ideal reference value.

On the basis of observations by Vigliani (1955a,b) and Pauwels (1955, 1965, 1980), Kummer (1962, 1963, 1972) formulated the interaction between mechanical stress and net bone remodeling as a mathematical function which took into account that below a certain stress value bone resorption does not progress any further, and that above a certain stress value the bone is destroyed by pathological resorption. Kummer (1971) used this function in the first recorded computer simulation of mechanoadaptive bone remodeling, in which the diameter of a one-dimensionally loaded cylinder of bone was adapted according to this mathematical rule and the effect of eccentric loading on its remodeling and shape was studied.

Frost (1964), however, emphasized that it is not actually the stress, but the strain in the bone that causes biological adaptation. Although the stress value and not the strain value was used in the formula developed by Kummer (1962, 1963, 1972), he and Pauwels (1955, 1965, 1980) assumed that it is ultimately the strain associated with these stress values that regulate the deposition and resorption of bone. In a linearly elastic tissue, stress and strain are closely coupled by the elastic modulus.

Cowin and Hegedus (1976) and Hegedus and Cowin (1976) put forward the first complete, constitutive, mathematical model of cortical bone remodeling: the so-called "theory of adaptive elasticity." This theory was formulated to be consistent with the laws of continuum mechanics and thermodynamics, the bone remodeling being regarded as the transmission of mass between the cortical bone and an (imaginary) fluid phase around it. The measure of this transmission was regulated by the strain of the material, and with departure from a known reference value (for which equilibrium of the remodeling was assumed), the stiffness of the material (characterized by the elastic modulus) was adapted. The stability of this formulation was dealt with in a later study (Cowin and Nachlinger 1978) and thereafter used to predict the internal remodeling process induced in diaphyseal bone by a medullary nail (Cowin and van Buskirk 1978). These studies took the distinction made by Frost (1964) between internal and external bone remodeling, by which the former (internal bone remodeling, in short: "remodeling") was related to the adaptation of the porosity and mineral content of the diaphyseal cortical bone, while the latter (surface remodeling or external remodeling, in short, "modeling") refers to an endosteal or periosteal bone deposition or resorption, with corresponding changes in the surface and cross-sectional form of the bone. A theory of this surface remodeling was developed in a further study, again exemplified by a medullary nail (Cowin and van Buskirk 1979).

Firoozbakhsh and Cowin (1981) demonstrated that in certain circumstances the theory of adaptive elasticity shows considerable similarity to the mathematical functions developed by Pauwels and Kummer. Cowin (1984) emphasized again that tissue strains – but not stresses – can be perceived because the latter can be calculated but not measured. He postulated that within a given range the bone reacts lazily or not at all to changes in mechanical loading, and it is only after exceeding or falling below critical thresholds that equilibrium is broken and effective mineral deposition or resorption takes place. However, because of the resulting mathematical complications, this nonlinear concept of a "lazy zone" or a "dead zone" was not taken into account when the original theory was formulated. It should be noted that this theory especially attempted to describe adaptive processes during changes in the mechanical loading, but not, however, to predict an optimal (that is to say, characterized by the maximum-minimum law) structure from the original homogeneous distribution of density.

The theory of adaptive elasticity of internal and external remodeling was finally implemented for the first time with the FE method by Hart et al. (1984a). They developed so-called three-dimensional remodeling FE models that allowed a refined mechanical analysis of animal experiments. These experiments provided data on the mechanobiological coupling by analyzing the quantitative relationship between controlled mechanical loads and the changes in density and shape of the bone which they produced. As noted above, in such an experiment the strains at the bone surface can be determined with strain gauges, but not the mechanical stresses within the material. Using the models of Hart et al. (1984a), it had now also become possible to investigate the relationship between internal and external remodeling.

In further studies, anatomically based models of tubular bone were derived from the animal experiments. A computer simulation of bone remodeling established a qualitative agreement with the alterations in the density and cross-sectional shape of the bone occurring during the experiments as a result of cyclic loading (Hart et al. 1984b).

2.3.2
Cellular Basis and Concept of Modern Bone Remodeling Algorithms

During bone remodeling, osteoblasts and osteoclasts act in close cooperation (coupling) and in a fixed sequence as so-called "basic multicellular units" (Frost 1964;

Parfitt 1982, 1994). There has been increasingly evidence in recent years that osteo-cytes and/or osteoblasts serve as mechanosensors for the bone tissue (Rodan et al. 1975; Somjen et al. 1980; Pead 1988; Skerry 1989; Marotti et al. 1990; Cowin et al. 1991; Lanyon 1993; Klein-Nulend et al. 1995a, 1997; Pitsillides et al. 1995a; Burger et al. 1996). It is thought that the osteocytes transmit signals through their processes to adjacent cells and thus affect the frequency of activation and/or the coupling of the osteoclasts and osteoblasts (Doty 1981; Menton et al. 1984; Palumbo et al. 1990; Marotti et al. 1990; Harrigan and Hamilton 1993; Weinbaum et al. 1994).

Duncan and Turner (1995) distinguished four phases of mechanotransduction:
- Mechanocoupling: the transduction of mechanical force applied to the tissue into a local mechanical signal perceived by the sensor cells (the osteocytes)
- Biochemical coupling: the transduction of a local mechanical signal into a biomechanical signal and, ultimately, changes in gene expression
- Transmission of signal: the communication of the biochemical response to neigh-boring cells that actually form or remove bone (the osteoblasts and osteoclasts)
- Effector cell response: the final tissue-level response, for example, proliferation, synthesis, and resorption of the bone matrix

Today it is widely believed that deformation of the bone causes fluid flow in the canalicular network, and that this, either through the shear stress in the cell processes or by means of streaming potentials, induces a biochemical response by the cells (e.g., Cowin 1991; Turner et al. 1994, 1995; Weinbaum et al. 1994; Zeng et al. 1994; Cowin et al. 1995; Duncan and Turner 1995; Smalt et al. 1996). It has been shown that osteoblasts in cell culture, either by fluid flow (Reich et al. 1990; Reich and Frangos 1991; Hung et al. 1995) or by direct mechanical strains (Brighton et al. 1991, 1992; Rawlinson et al. 1995; Pitsillides et al. 1995b) can be stimulated to synthesize a variety of intra- and extracellular messenger ions and molecules, such as, intracellular calcium, pro-staglandins (PGs), cyclical AMP, inositol phosphate, and nitric oxide (NO). However, recent experimental evidence suggests that fluid flow may comprise a more potent regulator than substrate deformation (Owan et al. 1997). Interestingly, cell cultures of skull bone are less reactive to mechanical stimuli than those of the long bones (Rawl-inson et al. 1995). Klein-Nulend et al. (1995a), and Burger et al. (1996) have demon-strated that pulsating fluid flow stimulates osteocytes (and also osteoblasts but to a lesser extent) to synthesize PGs and NO. Jacobs et al. (1998a) have compared the effect of steady, pulsatile, and oscillating fluid flow (at varying frequencies) on bone cells and conclude that physiological loading-induced fluid flow (which is oscillatory in nature and not steady or pulsatile) is a potentially important mechanical signal to the cell, but that experimentally oscillating flow is less stimulatory than steady or pulsatile flow. This suggests that mechanotransduction of oscillating flow occurs through fun-damentally different cellular mechanisms than steady or pulsatile flow. The frequency dependence of their results lead the authors to hypothesize that the response of bone cells to fluid flow is dependent on chemotransport effects, experiments having shown that physiological levels of mechanical loading applied to bones increase the levels of fluorescent tracers accumulated in the tissue (Knothe-Tate et al. 1998a,b). Recently a porous FE model of the rat tibia has been designed which predicts the load-induced fluid displacements as a function of load magnitude and loading frequency (Steck et al. 1999).

It has been shown that compression of their intercellular matrix brings about a change in volume and shape of the chondrocytes with concomitant effects on the cell membrane (Guilak et al. 1995). However, the deformation also concerns the cell nuclei (Guilak 1995), and that this effect is apparently mediated by the actin skeleton. The deformation of the nucleus has also been confirmed in compressed chondrocyte-agarose constructs (cell seeded systems) in the absence of a mechanically functional extracellular matrix (Knight et al. (1999). Compression of the cartilage has also been shown to drastically alter the morphology and packing of intracellular organelles (e.g., rough endoplasmic reticulum and trans-Golgi), with potential implications for the processing of matrix molecules, such as aggrecan (Grodzinsky et al. (1999).

Increased transcription and synthesis of β-integrins have been observed in bone cells after mechanical stimulation (Carvalho et al. 1995). The integrins couple the actin skeleton through the cell membrane with the extracellular matrix (Hynes 1992; Clover et al. 1992; Gailit et al. 1993). It has therefore been suggested that the cytoskeleton itself is important in the signal transduction process, one of the primary proteins that is activated as a result of such integrin engagement (Wang et al. 1993) being focal adhesion kinase (Schaller and Parsons 1994). There has been recent evidence that the integrins do not only act as mechanoreceptors in human bone cells, but that distinct integrin-mediated signaling pathways are activated by various frequencies of mechanical stimuli (Salter et al. 1997). Moreover, it has been suggested that bone and cartilage cells can regulate their mechanical microenvironment by varying their expression of integrins. Tenascin-C, an "antiadhesive" matrix protein that blocks integrin-mediated cell interactions with other components of the extracellular matrix, such as fibronectin and that can effectively weaken the cell connections with the matrix (Wang et al. 1993), has been shown to yield an increased expression in regions of high compressive strain in cartilage and tendons (Brand et al. 1997). Tenascin-C expression by bone cells is also enhanced in the early ostegenic response to loading (Webb et al. 1997). In osteoblasts the signal transduction pathway for mechanical activation of the osteopontin gene has been shown to depend on the integrity of the filaments (but not the microtubules), cytochalasin D (but not colchicine) being able to block the mechanically induced up-regulation of osteopontin expression (Toma et al. 1997).

In cell culture, fluid flow produces a rise in the intracellular calcium level of bone cells, probably mediated by inositol phosphate (Hung et al. 1995). These calcium signals can be passed on through "gap junctions" (consisting of connexin 43 proteins) to adjacent cells (Xia and Ferrier 1992), and the signal can therefore be effectively transmitted to other sites. Gap junctions have been shown to be increased in osteoblast-like cells after being subjected to mechanical strain, and this upregulation has been suggested to occur via the stimulation of connexin 43 mRNA expression (activation of the promotor by a load responsive gene element) and by messenger stabilization, and connexin 43 can thus be translated efficiently (Muccio et al. 1999). Intermittent hydrostatic compression of bone precursor cells (Klein-Nulend et al. 1995b) and cyclical strains of human osteoblasts (Neidlinger-Wilke et al. 1995) lead to an increase in production of transforming growth factor-β, a paracrine factor which contributes to the proliferation and differentiation of bone precursor cells. It is interesting to note that cells obtained from osteoporotic donors show neither strain-dependent transforming growth factor-β production nor raised proliferation rates (Neidlinger-Wilke et al. 1995). An increased gene expression for proliferation markers (c-*fos*), growth factors and bone matrix proteins was also found in vivo following the mechanical stimulation of periosteal cells (Raab-Cullen et al. 1994). Molecular biological investigations on bone precursor cells and osteoblasts have revealed that mechanical stimuli are necessary for maintaining the differentiation of the phenotype of bone cells (Roelofson et al. 1995). It has also been observed that the transcription and secretion of the matrix proteins is influenced by mechanical strain, without any hormonal induction being required (Harter et al. 1995). It has been shown that changes in osteoblastic gene expression begin immediately after the onset of the mechanical stimulation and last for 3–4 h. This has a selective effect on various cell products, the number of strain cycles apparently playing a more important role than their amplitude (Stanford et al. 1995). A time-dependent upregulation of prostaglandins (PGs) has been demonstrated, $PGF_{2\alpha}$ showing a rapid (5 min), PGE_2 a slower (10 min) and PGI_2 the slowest (30–60 min) response (Klein-Nulend et al. (1997). PGHS-2 mRNA expression (but not PGHS-1 mRNA expression) and PGHS activity are significantly increased by pulsatile fluid flow, PGHS being the major enzyme in the conversion of arachidonic acid to PGs. One of the earliest cellular responses reported, however, is that of nitric oxide (NO), the transcription and translation of constitutively expressed endothelial NOS isoform (but not that of neuronal NOS or inducible NOS) being clearly stimulated by fluid shear stress (Klein-Nulend et al. 1998).

The studies above have shown that bone cells react both in vivo and in vitro in many different ways in response to mechanical stimuli. Although some single elements of the mechanical transduction pathways are known, the understanding of the basic cellular processes involved is still far from complete. At the present time it seems likely that several regulatory systems exist side by side, and that they can interact in a relatively complex manner (Duncan and Turner 1995).

Apart from uncertainty about how the mechanical sensors in the bone work, it is also unclear by what criteria the optimization of bone is regulated. Possibilities include minimum weight, maximum resistance to fracture, and minimal stress and strain. It is also unknown what aspects of the mechanical environment are perceived by the cell. Is it the magnitude of the mechanical strains (Frost 1964; Cowin and Hegedus 1976), their rate (Lanyon 1984; Lanyon and Rubin 1984; Turner et al. 1995), their frequency (Turner et al. 1994; Rubin and McLeod 1995), the deformational energy (Fyhrie and Carter 1986; Huiskes et al. 1987), or microfractures (Carter et al. 1976; Martin and Burr 1982)? Are the mean values of decisive importance, or the peak values, or the gradients? Finally, it is not even clear whether bone does attain an optimal structure at all, or only an adequate structure (Huiskes and Hollister 1993). At the present time it is therefore not possible to offer, at the cellular level, a mathematical formulation of the laws by which mechanical stimuli lead to the adaptation of bone tissue.

The simulation of bone remodeling as a function of the mechanical loading by means of numerical methods (such as FE method) requires, however, an exact formulation of the operations by which mechanical signals are transformed into a biological response: so-called "algorithms." The bone-remodeling algorithms presently available attempt to imitate the basic biological remodeling processes "phenomenologically" (Huiskes 1995). This means that the actual cellular mechanisms of the mechanotransduction are not in fact simulated, but that one attempts to predict the effect of the mechanically induced bone remodeling on interesting skeletal regions accurately. This has led to criticism of the approach (e.g., Bertram and Swartz 1991;

Currey 1995). It has been argued that this kind of simulation corresponds to "curve fitting" at a high technological level rather than to the actual testing of hypotheses. However, as Huiskes (1995) has noted, such an empirical method has enjoyed a long tradition in physics and was inspired by Newton himself. In this way the validity of the bone-remodeling algorithms is confirmed by their "predictive value," that is to say, by their ability to make accurate predictions which agree with currently available physiological and clinical observations, as well as with the results of specific animal experiments. Exact knowledge of the basic biological regulatory processes is certainly desirable, but not an absolute necessity in this context.

In spite of certain nuances in their formulation, contemporary bone-remodeling algorithms obey a common fundamental principle: bone can be interpreted as a "mechanostat," in much the same way that a thermostat (bound into a feedback loop) regulates the temperature set to a predetermined value (Frost 1987). The mechanostat adapts itself in terms of bone shape, density and structure to the corresponding mechanical requirements. A particular aspect of the mechanical conditions serves as a feedback value or signal which is perceived by a sensor (probably the osteocytes) and compared with a preset reference value. Any departure of this signal from the reference value produces a stimulus (modulated by genetic, hormonal or metabolic factors) which causes the "actors" (the osteoblasts and osteoclasts) to have a regulatory effect on the local deposition or resorption of bone tissue. This gives rise to a new form of the bone (external remodeling, surface remodeling or modeling) and to a different internal structure (internal remodeling or remodeling). This in turn leads to a change in the mechanical material properties, which in combination with changes in the geometric relationships brings about a different distribution of the mechanical signals within the bone (Fig. 4). In order to perform a computer simulation of functional bone remodeling, the following questions must be answered:

- Where is the sensor located?
- What is the mechanical signal that is perceived by the sensor, and to what aspect of this signal does the sensor react?
- How large is the reference value with which the signal is compared?
- By what factors is the stimulus (that results from a difference of the mechanical signal from the reference value) modulated, and what is the quantitative relationship between the magnitude of the stimulus and the extent of the bone remodeling?
- How is the remodeling converted into a change in the external form and the internal structure and density of the bone?
- What is the relationship between the structure and density of the bone and the mechanical properties?

The mathematical formulations of these relationships (Fig. 4) constitute bone-remodeling algorithms. To what extent an alteration in the geometry and the mechanical material properties brings about another distribution of the mechanical signals can be analyzed by FE method. The bone-remodeling algorithm serves to actualize a corresponding FE model by adapting its external shape and the material properties (preprocessing; see above). Under these changed conditions a new distribution of the mechanical signal can be calculated by means of FE analysis (processing and postprocessing). In this way the feedback loop is closed (Fig. 4). The iterative process is repeated until the external shape and the internal structure and density of the body show no further significant change. At this point the bone-modeling simulation is said

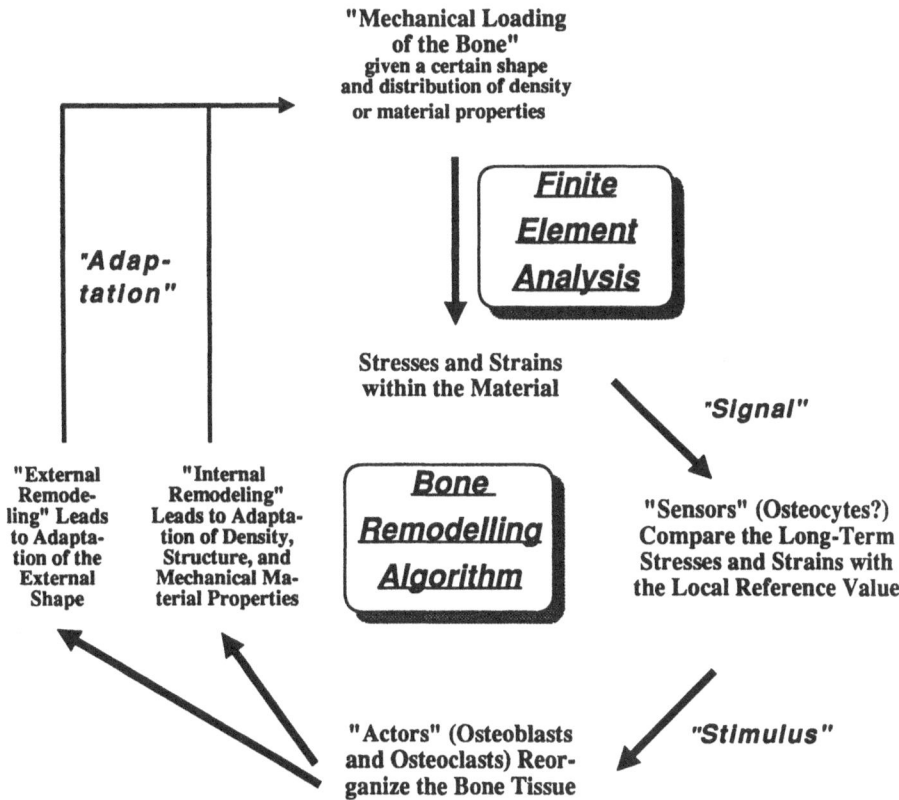

Fig. 4. Concept of modern mechanoadaptive bone remodeling simulations

to have converged. A structure emerges from the simulation which is ideally adapted to the given mechanical loading, a change in the structure being expected only after subjecting it to a different type of loading.

The differences between the various theories and algorithms consist essentially in the differing formulations of the points 1 to 6 listed above. In the following section, current homeostatic and time-dependent formulations of bone remodeling, their application in remodeling simulations and the problem of numerical instability will be considered.

2.3.3
Homeostatic and Time-Dependent Formulations, Experimental Validation, and Numerical Instability

The need for a mathematical theory of bone-remodeling is related on the one hand to the interest in acquiring a basic understanding of the form-function relationship of bone tissue, and on the other hand to the need to predict the long-term effect of osteosynthetic implants or endoprostheses on their immediate neighborhood (osteointegration). Numerical simulation of mechanically induced bone-remodeling can be applied for optimizing the design and the material properties of prostheses, thus increasing the success rate in treating patients.

Carter (1984) reviewed the conceptual framework of bone-remodeling algorithms, referring to cortical bone. The goal of mechanoadaptive bone remodeling is, in his view, to produce a resistant structure that can withstand mechanical impact loading, and, secondly, to repair the microfractures occurring during normal physiological loading over time to prevent fatigue fracture. In this context it is possible that the bone remodeling process is regulated by several complementary mechanical signals, one of which can be a microfracture. The reference values of these signals are unequally distributed throughout the skeletal system; the bones of the skull, for instance, requiring less mechanical stimulation for their preservation than those of the extremities (site-specific response). The relationship between the mechanical stimulus and the biological response is nonlinear (concept of the "dead" or "lazy" zones; Fig. 5). This is to say, that the bone reacts either very little to small departures from the required equilibrium, or not at all. In his view it is possible that bone resorption and deposition are controlled by different mechanisms, and that adaptation to reduced demands is more rapid than that to increased loading (Fig. 5). This response also depends on the age or on the maturity of the bone (time-specific response).

Building on this conceptual framework, Fyhrie and Carter (1986) developed a mathematical theory to predict both the adaptation of the trabecular bone density and that of trabecular orientation to a distinct mechanical loading condition. Whereas the density is regulated by a directionless scalar quantity of mechanical stress and strain, the trabecular orientation depends upon the direction of the principal stresses. Two different optimization criteria were studied: the SED and a so-called effective stress that is calculated on the basis of a failure criterion (the Tsai-Wu criterion). The former leads to an optimal distribution of mechanical stiffness and the latter to an optimal distribution of strength. These two formulations take on a similar mathematical form and lead to comparable results with regard to both density and trabecular orientation. It has been observed that the von Mises stress, originally formulated as a failure criterion for metals, is unsuitable as a feedback signal.

Fig. 5. Possible relationships between the rate of bone remodeling and the magnitude of the mechanical stimulus (stimulus=magnitude of the mechanical signal minus reference value)

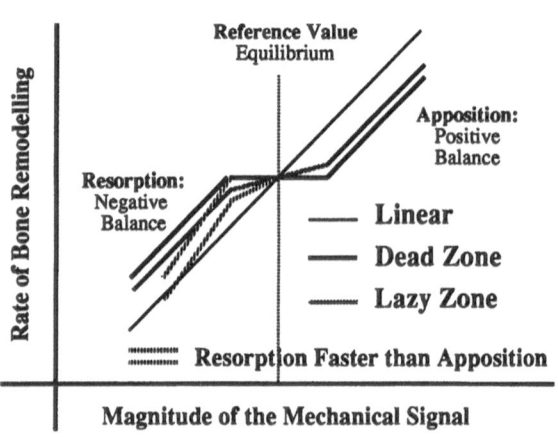

In further developing this theory, Carter (1987) and Carter et al. (1991) proposed a concept by which a whole range of biological phenomena of the connective tissues could be related to mechanical factors, beginning with differentiation, morphogenesis, and growth, through regeneration and functional adaptation, to degenerative changes. In a two-dimensional iterative computer simulation of functional bone remodeling, the typical features of the morphology of the proximal femur – that is to say, the development of cortical diaphyseal bone, the marrow cavity and the characteristic density distribution in the epiphysis and metaphysis – were obtained from an initial homogeneous density distribution. In a further study (Carter et al. 1987c), various optimization criteria were analyzed, including an effective stress (on the basis of a failure criterion), the fatigue damage, and the SED. It was established that, because of the mathematical similarity, no particular formulation offers any objective advantage. A later publication (Carter et al. 1989) undertook further computer simulations of bone remodeling in the proximal femur, in which the mechanical signal was taken as dependent upon both the magnitude of cyclical loads and the number of loading cycles appearing during the stance phase of gait. In order to simulate this complex mechanical situation adequately, several load cases (three joint forces, and corresponding muscle forces acting at the trochanter) were applied. The total loading was obtained from nonlinear superimposition of all load cycles associated with these load cases. An equal distribution of the reference values throughout the femur (non-site-specific response) and a linear relationship between the stimulus and the rate of remodeling were assumed (Fig. 5). After about three iterations the typical architecture of the proximal femur was predicted. In the course of further iterations, however, unphysiological patterns appeared in which single elements of the FE mesh tended to assume either the highest or lowest possible density value. Since the effective normal stress (from which the orientation of the trabeculae was predicted) also accounted for various load cases, it follows that the trabeculae do not necessarily have to align at 90° angles. According to Carter, the predicted trabecular architecture of the proximal femur is actually more similar to that described by von Meyer (1867) than the corrected illustration by Wolff (1892). Fyhrie and Carter (1990) were also able to predict the essential characteristics of the physiological density distribution, simulating a typical load case in a three-dimensional model of the femoral head.

Huiskes et al. (1987) developed another mathematical theory of bone remodeling and a two-dimensional FE model of the proximal femur, the primary aim of which was to predict the long-term processes which follow the implantation of an artificial hip joint. The SED was used as a mechanical feedback signal and a reference value defined for every point, the latter being derived from an analysis of the model before the introduction of the endoprosthesis (site specific approach). The interdependence between the mechanical stimulus and the rate of remodeling was formulated as trilinear, i.e., including a "dead zone" (Fig. 5). Separate simulations of the internal remodeling (changes in the material properties/the elastic modulus) and the external remodeling (changes in the form/relocation of the nodal points on the surfaces of the model) were performed, and only one load case taken into account. In view of the site-specific reference values, Huiskes et al. (1987) differentiated between their formulation and Carter's "theory of self-optimization," calling it the "theory of adaptive elasticity," following Cowin and Hegedus (1976). According to Huiskes, however, the two theories agree in that each point is optimized by itself (local self-optimization) and not the whole structure (whole structure optimization). The latter could be guaranteed, for instance, by an evolutionarily successful genetic program. The authors also noted that Carter's three load cases can be represented as a single force when looking at diaphyseal bone and using site-specific reference values, as long as axial compression and bending of the femoral shaft are included. The stiffness and fixation of the implant were found to have a sustained effect on the extent of the bone resorption because of so-called "stress shielding." This means that the load, which is normally transmitted by the bone tissue, is diverted distally by the prosthesis, with the result that the proximal part of the bone undergoes atrophy. These results confirm both clinical observation (Engh and Bobyn 1984) and the results of animal experiments (Turner et al. 1986).

In the homeostatic formulations of bone remodeling, the local material properties of the tissue were adapted at each iteration step, so that the difference between the mechanical signal and the local reference value was completely eliminated. This neglects the fact that the osteoblasts and osteoclasts require a certain amount of time to remodel bone. It is therefore only to a limited extent that these theories represent the real course of events.

The so-called time-dependent (evolutionary) formulation of bone remodeling takes into account the free surface available for the osteoblasts and osteoclasts to act. Basing their work on the suggestions of Carter and coworkers, Beaupre et al. (1990a,b) put forward a time-dependent formulation of bone remodeling. This "Stanford formulation" of bone-remodeling is the basis of the simulations described in the present work.

The theoretical development (Beaupre et al. 1990a) includes internal and external bone remodeling ("remodeling" and "modeling"; Frost 1964) in a single theory. The definition of the mechanical feedback signal addresses two fundamental problems. Mechanical loading and bone remodeling occupy different time scales; the former takes place cyclically and typically at intervals of seconds while the latter occupies weeks, months, or even years. Determining the relationship between the two processes requires considering both the integral of numerous load cycles acting within a given time period and the effect of various load cases. A load case is determined by a specific magnitude and direction of force and is in each case bound up with a particular number of daily loading cycles. Furthermore, a mechanical signal can act only in the mineralized bone and be registered there by the hypothetical sensors, but not in the intertrabecular marrow spaces. The mechanical loading at the level of the mineralized tissue cannot, however, be derived directly from the FE analysis, since with this approach the value is calculated at the continuum level. An exception to this are very recent models in which the trabeculae themselves are represented (e.g., Beaupre and Hayes 1985; Hollister et al. 1991; Müller and Rüegsegger 1995; van Rietbergen et al. 1995). This has been possible, however, only for small tissue samples and only very recently for entire bones (van Rietbergen et al. 1999; Ulrich et al. 1999).

Beaupre et al. (1990) calculated the relevant mechanical feedback signal as the average daily tissue level stress at the level of the mineralized bone tissue (ψ_b). This was defined as:

$$\psi b = \left(\sum_{i=1}^{N} n_i \sigma_{b_i}^{m} \right)^{1/m} \tag{1}$$

in which N is the number of load cases, n_i the average number of daily load cycles in load case i, σ_{bi} the current stress intensity at the level of the mineralized bone tissue (i.e., not at the level of a continuum) in load case i, and m an empirical constant. The constant m must be understood as a weighting factor which decides in what proportion the magnitude and the number of loading cycles contribute to the total stress. Whalen et al. (1988) determined a value for m of between 3 and 8, which means that the magnitude of the loads is more important than the number of cycles. In other words, loading cycles of large amplitude have a more sustained effect on bone remodeling. The current stress intensity at the level of the mineralized tissue σ_{bi} can be calculated from the corresponding value in the continuum model on the basis of the interrelationship given by Carter and Hayes (1977) and Gibson (1985) between the local bone strength and density:

$$\sigma_{b_i} = \left(\frac{\rho_c}{\rho}\right)^2 \sigma_{c_i} \tag{2}$$

where ρ_c is the density of the cortical bone (which is here equal to the density of the mineralized tissue), ρ the density of the trabecular bone (i.e., the mass of the mineralized tissue per unit volume), and σ_{Ci} the stress intensity at the level of the continuum in load case i. The stress value at the level of the continuum was determined in the FE analysis and is given by:

$$\sigma_{c_i} = \sqrt{2EU} \tag{3}$$

where E represents the elastic modulus and U the SED. U is given by:

$$U = \frac{\sigma : \varepsilon}{2} \tag{4}$$

where σ is the stress tensor and ε the strain tensor.

Using a combination of the Eqs. 1–4, the desired signal ψ_b can be obtained from a FE analysis. The so-called "tissue level remodeling error" represents a figure that indicates to what extent the density of the bone differs from the desired remodeling equilibrium. It is calculated from the difference between ψ_b and a given reference value, and represents the actual stimulus for the remodeling process. The theoretical development of the theory (Beaupre et al. 1990a) considers the reference value to be dependent upon the skeletal location, genotype, metabolic conditions, and biochemical interactions with the surrounding tissues. The associated computer simulation (Beaupre et al. 1990b), however, used a simplified uniform reference value of 50 MPa. This was derived from experimental measurements made with strain gauges in an animal model (Rubin and Lanyon 1984) and also on a human volunteer (Lanyon et al. 1975). It follows that:

$$e = \psi b - 50 \text{ MPa} \tag{5}$$

where e is the so-called "remodeling error," that is to say, the actual stimulus for the bone remodeling.

A nonlinear relationship between the stimulus and rate of remodeling was used which assumes a lazy zone around the equilibrium conditions and, according to Parfitt (1983) and Frost (1986), shows steeper slopes when it falls short of the reference value (bone resorption) than when it exceeds it (bone deposition) (Fig. 5). The theory proposes a piecewise linear relationship consisting of four sections showing different slopes. The simulation, however, used a simplified trilinear relationship with a "dead zone" equal to 20% of the reference value (Fig. 5) and an identical slope of 0.02 µg/day for each MPa per day of mechanical stimulus for both bone deposition and resorption. The value resulting from the nonlinear mathematical relationship was then used for the calculation of the remodeling rate at the endosteal and periosteal surfaces of the bone, since here the osteoblasts and osteoclasts have almost unlimited spatial access. A computer simulation of bone remodeling at the proximal femur (Beaupre et al. 1990b), however, analyzed only the internal remodeling of the bone. A limited surface area per unit bone volume is assumed to be available for the osteoblasts and osteoclasts to act. This characteristic free surface for each density was designated the "specific surface." According to Martin (1984) the relationship between the specific surface or "bone area surface density" and the apparent density of the tissue can be expressed by a fifth order polynomial function which has a maximum at the mean density value and declines from this in either direction. By means of a histomorphometric equation put forward by Frost (1983) it was then possible to calculate the change in density within a given time interval (usually 10 days) from the local remodeling rate. From the resulting density (ρ), the relationships given by Carter and Hayes (1977) and Gibson (1985) were then used to calculate the elastic modulus (E) of the bone tissue, using the equations:

$$E = 2014\, \rho^{2.5} \tag{6}$$

when $\rho < 1.2$ g/cm^3, and:

$$E = 1763\, \rho^{3.3} \tag{7}$$

when $\rho > 1.2$ g/cm^3 (E, elastic, or Young's, modulus; ρ, apparent bone density).

In the computer simulation (Beaupre et al. 1990b) a two-dimensional FE model of the proximal femur was started with an initial homogeneous elastic modulus of 500 MPa and a Poisson ratio of 0.2. Following the simplified procedure described above, the mechanical signal, stimulus, and deposition and resorption rates were determined for each element. Thirty iterations were performed for a total of 247.5 days. When the resulting changes in density were below 0.02 g/cm^3 per time step the simulation was accepted as having converged. The load magnitude was then reduced by 30% and the simulation restarted. After renewed convergence the femur was again loaded with the original initial force. In the first step, the development of a physiological (inhomogeneous) density distribution out of the initial homogeneously distributed density (or material stiffness) was observed. Comparison with a simulation lacking a "dead zone" (linear relationship; Fig. 5) produced less realistic results. A reduction in loading led to bone atrophy that was only partly restored by reintroducing the initial load magnitude for the same period of time.

Weinans et al. (1989) developed a very similar time-dependent theory with a uniform distribution of the reference values (not site-specific). This formulation, known as the "Nijmegen theory," was also based on Martin's suggestions (1972) regarding the specific surface available to the osteoblasts and osteoclasts. Weinans et al. (1989) demonstrated that the simulation at the proximal femur always led to convergence, and that this was relatively independent of the choice of the initial density distribution. Although the results for different initial conditions were not identical, they were very similar.

In a later work, Weinans et al. (1993) undertook a three-dimensional analysis in direct imitation of the implantation of uncemented hip prostheses in experiments on beagle dogs (Turner et al. 1986; Sumner et al. 1992). The SED served as a mechanical feedback signal; unlike Beaupre's simulation, however, this simulation assumed an inhomogeneous site-specific reference value, which was derived by the distribution of the SED in an FE analysis of the contralateral (unoperated) side. This variant of bone-remodeling simulation, designated as conservative, reacted in a less sensitive manner to the selected load application than the non-site-specific formulation. Both external (geometric) and internal (density) remodeling were simulated at the same time, and the time interval between two iterations was chosen in such a way that a defined maximum change in density was not surpassed. Under these conditions there was relatively good agreement between the prediction of the empirical model and the results of the animal experiments, both with regard to the changes in the cross-sectional form of the bone and in its density. It was concluded that the morphological changes around the hip prosthesis can be explained entirely by the natural adaptation of the bone to the changed mechanical situation, and that the algorithm developed has a high predictive capacity. It was also noted, however, that simplifications of the model and uncertainty in the choice of some model parameters account for some differences between the prediction and the experimental findings, and that further development of the theory is therefore necessary. In a very similar work van Rietbergen et al. (1993) compared the model predictions and the results of animal experiments for noncoated "press-fit" endoprostheses. An acceptable agreement between experiment and computer simulation was again established, which lay within the 95% confidence interval for the results of the animal experiments.

An important problem with these simulations was, however, that with a large number of iterations they had a tendency to produce discontinuous density patterns. This means that immediately adjacent elements take on either the minimal or the maximal possible density value (Carter et al. 1989; Weinans et al. 1989, 1992; Harrigan and Hamilton 1992). This phenomenon was attributed to numerical instability, with Jacobs et al. (1995) distinguishing two varieties of this phenomenon: (a) instability distant from the site of the load application (far-field) and (b) instability in its proximity (near-field). This distinction is important, since only the occurrence of the former allows a realistic prediction of the physiological density pattern, such as the form of the diaphyseal cortex and the marrow cavity. The second phenomenon, however, introduced "checker-board patterns" in the epiphysis, and these were neither physiologically plausible nor compatible with the continuum assumption of the FE analysis. A decisive question was, whether a single remodeling theory (and thus a single biological adaptation mechanism) can explain the cortical (discontinuous) structure of the diaphysis as an adaptation to a relatively uniform mechanical loading, and the trabecular (continuous) structure of the epiphysis as an adaptation to a relatively complex and variable loading. It must be noted that, fundamentally speaking, the trabeculae in the epiphyses and metaphyses are themselves naturally discontinuous morphological structures. As discussed above, it is nevertheless not yet possible to simulate the remodeling of entire human bones down to the level of the trabeculae. In the present formulations, both the mechanical signals and the biological response must therefore be calculated at the level of the continuum, and the problem of instability must be solved at this level.

Harrigan et al. (1988) have described the limited validity of the continuum assumptions for trabecular bone. They concluded that this can lead to false results at biological interfaces and under stress gradients of more than 20%–30% within a region of three to five trabeculae. Harrigan and Hamilton (1992) analyzed the cause of the instability of current SED-based bone remodeling algorithms by means of analytical and numerical methods. They came to the conclusion that the instability is principally a local effect that can be traced back to the ratio between n and m in the description of trabecular bone, where n characterizes the exponential relationship between the elastic modulus and the density, and m that between the strength and the density (see above; 2.5, Carter and Hayes 1977; or 3.3, Gibson et al. 1985). The "tissue level stress stimulus" is calculated based on these two values, and, according to Harrigan and Hamilton (1992), m must be greater than n to guarantee stability under all conditions, but this relationship has not been experimentally observed in trabecular bone. As long as m is less than n, the increase in density in one element can produce a rise in local stimulation (attraction of the SED) and thus lead to a further increase in density until the maximum possible value has been reached. On the other hand, a slight reduction in the density can bring about the opposite effect, since the corresponding element is shielded from mechanical stress. The authors therefore concluded that instability of current bone remodeling simulations is due at least partly to the mathematical formulation of bone remodeling algorithms, and not to the approximation procedure involved in numerical methods.

Weinans et al. (1992) also described the phenomenon of a "checker-board" (they called it a "patchwork") pattern as an intrinsic problem of bone-modeling simulations. They found that, in the relationship between the elastic modulus and the density, only exponents $n/m<1$ lead to a stable solution. With values $n/m>1$ (according to Rice et al. 1988 and Hodgskinson and Currey 1990, realistic values lie between 2 and 3) the continuous density distribution reached after several iterations (at which point almost all elements are in equilibrium) constitutes an unstable saddle-point. Thereafter the system diverges in further iterations to a stable discontinuous system. Self-accelerating, positive local feedback mechanisms lead to various solutions. A large number of the elements took either the highest (2.0 g/cm^3) or the lowest (0.2 g/cm^3) possible value. This phenomenon was independent of the nature of the elements used and of the refinement of the FE mesh. The authors regarded the bone as a chaotically formed, fractal structure, with optimal mechanical resistance but the minimum possible mass. They explained the divergence by the concept of self-optimization (as against a general optimization, Kuiper et al. 1991), the total mass of the discontinuous solution being less than that of the continuous, homogeneous distribution. The authors remarked that the final discontinuous configuration of the model may describe the actual trabecular structure of the bone, but that the solution calculated on the basis of interpolations (carried out by the postprocessor of the FE program) comes very near to the macroscopic, apparent density distribution of the real femur.

Jacobs et al. (1995) remarked, however, that the analyses reported in these two studies represented an accurate model of the actual condition found in the shaft of a long bone, but that in the case of complex mechanical loading (such as may appear in the epiphysis) the reported observations may not be entirely valid. This is because the local stress and strain values are affected in a different way by the general density distribution, and an analytical solution (as forwarded in these analyses) is therefore inappropriate. It was suggested that (near-field) discontinuities are brought about by an interaction of the bone remodeling equations with the discrete nature of the FE method. The authors therefore developed a new implementation technique for the remodeling algorithms with the FE method.

Traditionally the stress and strain values for the single elements have been calculated from the displacement of the nodes, and a homogeneous distribution of the density within each element of the FE mesh has been assumed (element-based type of procedure). However, as an alternative to this, the stress and strain values of the elements can be projected on to the adjacent nodal points, and then the density values for the points of the mesh can be predicted by the above mathematical equations. The density values of the elements, which are necessary for assembling the stiffness matrix and carrying out the following iteration, must then be interpolated from the values calculated at the nodal points. In this way, discontinuities of the stress and strain values at the borders of the elements are smoothed out. Jacobs et al. (1995) showed in a two-dimensional FE model of the proximal femur that with this formulation the desired "far-field" discontinuities, which allow the formation of the marrow cavities and diaphyseal cortex, remained. On the other hand, the undesirable "near-field" discontinuities at the epiphyses and metaphyses vanished. It was thus possible to predict both continuous and discontinuous structures on the basis of a single SED-based coupling system. This does not necessarily imply that this mechanism is actually valid at the cellular level, but it does substantiate the high predictive capacity of the formulated bone modeling algorithm.

[It is interesting to note that the implementation modification proposed by Jacobs et al. (1995) is very similar to one proposed by Diaz and Sigmund (1995) in a completely independent context. One of the classic problems in the field of structural optimization is to design a mechanical part or device to maximize its stiffness with a given amount of material. The typical approach is to adjust the material density in the elements of a FE model with the goal of increasing structural stiffness. Such algorithms posses many similarities with bone remodeling algorithms and can also exhibit checkerboard patterns in certain situations. In the field of structural optimization several smoothing techniques (known as filtering algorithms of which the nodal technique of Jacobs may be considered one variation) have now become standard procedures.]

Since the subchondral bone lies particularly close to the site of the load application, where it is subjected to complex and highly variable loading conditions, one may expect artifacts due to numerical instability with a traditional element-based implementation. For this reason, in the present study, we implement Beaupre's theory of bone remodeling with the nodal technique introduced by Jacobs (1994) and Jacobs et al. (1995).

3 Materials and Methods

3.1
Idealized Model of Concave Joint Incongruity

An idealized model of "concave" joint incongruity was designed (i.e., a joint in which the concave component is deeper than required for an exact fit with the convex joint component) based on quantitative measurements in the humeroulnar joint (Eckstein et al. 1993, 1995a). In this example the concave component (the trochlear notch) was constructed as an ellipse, its long axis exceeding the radius of the convex component (the trochlea of the humerus) by 10% (Fig. 6). The two-dimensional "plain-stress" FE model had a width of 20 mm and consisted of 2017 bilinear elements.

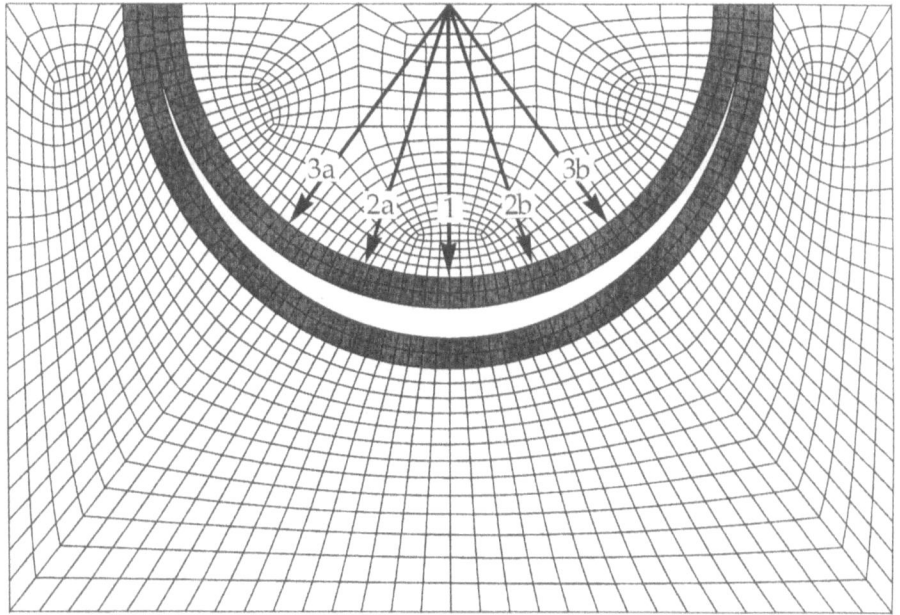

Fig. 6. FE model of an idealized "concavely" incongruous joint with a 10% deeper joint socket; *1–3*, five simulated load cases; *1–3*, widespread loading; *1*, central loading; *3a*, *3b*, bicentric loading; *1*, *2a*, *3a*, asymmetric loading; *dark elements*, cartilage; *light elements*, bone. With permission from Eckstein F, Jacobs CR, Merz BR (1997) Mechanobiological adaptation of subchondral bone as a function of joint incongruity and loading. Med Eng Phys 19:720–728 (Copyright Elsevier Science 1997)

Two 1-mm cartilage layers were modeled (Kurrat and Oberländer 1981; Oberländer and Kurrat 1982; Schenck et al. 1994; Milz et al. 1997; Springer et al. 1998), each being represented by two rows of elements. For the articular cartilage an infinitesimal isotropic linear-elastic material model was selected, describing the initial deformational behavior of the tissue under short-term cyclic loading. This model assumes that the stresses and strains stand in a linear relationship to one another, that the material properties are independent of the direction of the applied force, and that the strains do not exceed a value of about 20%. Based on the investigations of Kim et al. (1995), in which the "dynamic stiffness" of the cartilage was determined for oscillating loading in unconfined material samples and those of Shephard and Seedholm (1996), in which the stiffness was measured for a short loading cycle with a physiologically realistic duration of about 150 ms, an elastic modulus for cartilage of 15 MPa was selected in our study. Because of the biphasic material properties of cartilage its stiffness depends upon the rate and duration of the loading, and the elastic modulus can only approximate the observed behavior of the material. The actual value depends upon the time period elapsing between the beginning of the load application and the value measured, and it can therefore not be unambiguously determined. For this reason additional simulations were performed for elastic modules of 5 and 25 MPa to evaluate the sensitivity of the simulation to various material properties of the cartilage. Since articular cartilage behaves as an incompressible elastic material under short-term loading (Hayes et al. 1972), a Poisson ratio of 0.499 was chosen. This means that under loading the total volume (but not the local thickness) of the cartilage remains virtually the same.

Since with concave incongruity the position and extension of the contact areas between the joint components vary with the direction and magnitude of the joint reaction force, these cannot be defined a priori. However, just as the FE method can be used for the analysis of stresses and strains in the tissue, it can also be employed to solve such nonlinear contact problems. Assuming a frictionless surface (which has been shown to be true for cartilage; Fung 1993; Ateshian 1997; Wang and Ateshian 1997), the contact between the joint surfaces can be simulated by so-called "slide-line elements" and may be calculated for every single load cycle by the FE program.

An isotropic linear elastic material model has also been used for the bone. The static analysis of the mechanical stresses and strains in the model was based on a homogeneous density distribution of 0.5 g/cm^3, and the bone remodeling simulations were also started from this homogeneous value. To assess the dependence of the final configuration on initial conditions additional simulations were started from a minimal (0.2 g/cm^3) and a maximal (2.0 g/cm^3, i.e., cortical bone) density, and from a density distribution obtained from random numbers, both with and without taking into account a "dead zone" (Fig. 5). All subsequent simulations were undertaken with a dead zone, which amounted to 20% of the reference value (10 MPa). The bone remodeling theory of Beaupre et al. (1990a,b) was implemented by means of a nodal technique (Jacobs 1994; Jacobs et al. 1995a) with the FE program ABACUS 5.4, and the calculation carried out on a DEC 3000/900 workstation. The remodeling process was simulated in 30 iterations of 10 days each, corresponding to a total period of 300 days. After this time no further significant changes in the density (<0.05 g/cm^3) were observed.

The nodes at the basis of the model were locked in the x- and y-directions, but both sides of the joint socket were free to bend laterally. Using estimations of the joint forces for various angles of flexion in the elbow (An et al. 1984; Morrey 1992; Donkers et al. 1993), five load cases were selected. In the first case the head was pressed centrally into the socket, in two cases 22.5° laterally (Fig. 6), and in two cases 45° laterally (Fig. 6). Various combinations of these load cases were used in the bone modeling simulations: (a) all five cases (widespread loading); (b) only case 1 (central loading); (c) cases 3a and 3b (bicentric loading) and (d) cases 1, 2a, and 3a (asymmetrical loading). Assuming that with strenuous activity the forces at the elbow joint can reach 0.4 times the body weight, and that they are more or less equally distributed between the humeroradial and humeroulnar joints (Halls and Travill 1964), forces of 140 N (heavy loading), 70 N (moderate loading), and 35 N (light loading) were applied. The number of daily load cycles was varied between 1000, 3000, and 10,000.

3.2
Idealized Comparative Geometric Models

To assess selectively the effect of incongruity on the load transmission and the remodeling process, four geometric models were constructed for comparison, each with a different type of incongruity. To reflect the geometric configuration of the humeroradial joint at particular angles of flexion, an elliptical socket was in one case constructed with the major axis exceeding that of the head by 10% (Fig. 7a). Correspond-

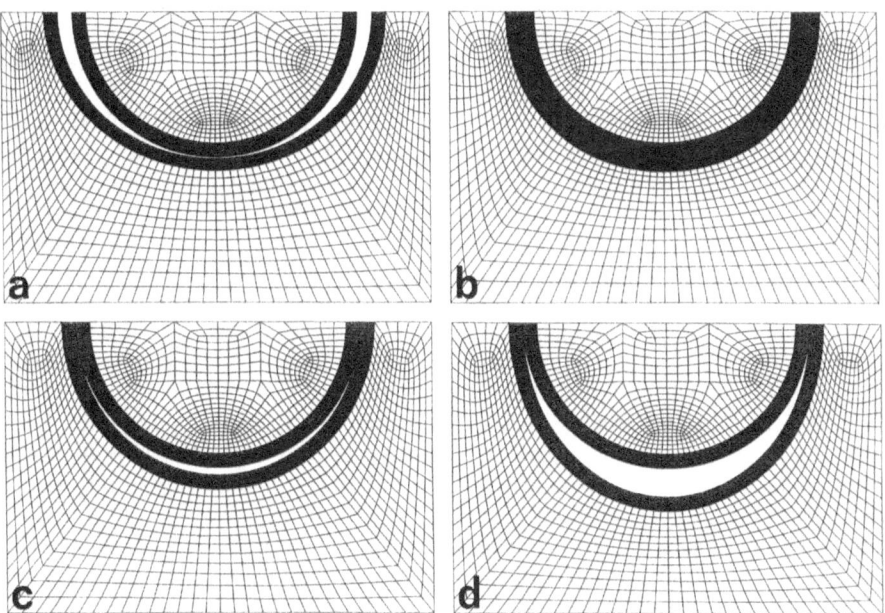

Fig. 7a–d. Comparative idealized FE models with various joint shapes. **a** "Convex" incongruity (10% wider socket). **b** Congruity. **c** "Concave" incongruity (5% deeper socket). **d** "Concave" incongruity (20% deeper socket). With permission from Eckstein F, Jacobs CR, Merz BR (1997) Mechanobiological adaptation of subchondral bone as a function of joint incongruity and loading. Med Eng Phys 19:720–728 (Copyright Elsevier Science 1997)

ing to the conditions in the humeroradial joint at other flexion angles, and in exceptional cases to those in the humeroulnar joint, we also constructed a congruous case (Fig. 7b). Because of the quantitative variations in the concave incongruity of the humeroulnar joint, additional models with 5% and 20% deeper sockets were designed (Fig. 7c, d).

All other parameters were chosen exactly as in the case of the idealized model with a 10% concave joint incongruity. To obtain an overall comparison of the predictions of the various models under various loading conditions, the density of the subchondral bone in the model (represented by the row of nodes along the bone-cartilage interface of the concave joint component) is presented graphically as a function of the distance to the center of the articular surface.

3.3
Anatomically Based Model
of the Humeroulnar Joint

Whereas the idealized models described previously are suitable for obtaining principal information and for performing parametric investigations, the anatomically based model of the humeroulnar joint served to test and validate the results obtained by the former models for the conditions of one specific joint. The objective is to simulate the real anatomical relationships and boundary conditions as closely as possible, and to verify the results experimentally.

The model was based on a fresh specimen of the humeroulnar joint from a 33-year-old man without visible signs of pathological change and with a typical transversely divided articular surface of the trochlear notch (Tillmann 1971, 1978). The two-dimensional FE model was constructed from a sagittal section through the joint passing along the longitudinal ridge of the ulna (Fig. 1), and continuing along the course of the shaft distally (Figs. 1, 8). To that end, a sagittal CT section (resolution $2 \times 0.2 \times 0.2$ mm^3) was first prepared with a Somatom-Plus-4 scanner (Siemens, Erlangen, Germany). A phantom was used with 0 and 200 mg/cm^3 hydroxyapatite equivalents and the density determined at intervals of 200 HU (Fig. 9).

Following removal of the soft tissues, the capsule and the ligaments, the joint components were positioned in special devices in a material testing machine (Zwick 1445, Ulm, Germany) (Fig. 10). A tension jig was attached to the olecranon through which a force could be applied in the proximal direction to simulate the action of the triceps brachii. The ulna shaft rested on a transverse rod, its position determining the flexion angle of the elbow joint. The shaft could move freely in all directions on the rod (translation and rotation), so that the ulnar joint surface was brought into natural contact with the humerus without relevant constraining forces.

As a measure of the anatomical incongruity (or the relative difference in shape of the articular surfaces) the width of the joint space and the contact area was determined in the same way as in previous studies (Eckstein et al. 1993, 1994b, 1995a) by means of a polyether casting material (Permadyne, ESPE, Seefeld, Germany) with a load of 25 N. This force was necessary to displace the casting material and to bring the joint components into contact. The thickness of the cast was then measured by a special device (IDS 543, Mitutoyo, Neuss, Germany; technical accuracy ±0.02 mm) at defined points represented by cylindrical coordinates. The ten points coinciding with

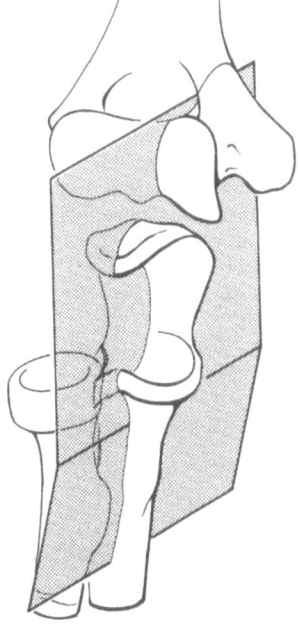

Fig. 8. Plane of the two-dimensional, anatomically based model of the humeroulnar joint. The model follows proximally the main sagittal ridge of the trochlear notch, and distally the shaft of the ulna. With permission from Merz B, Eckstein F, Hillebrand S, Putz R (1997) Mechanical implications of humero-ulnar incongruity – finite element analysis and experiment. J Biomech 30:713–721 (Copyright Elsevier Science 1997)

Fig. 9. Sagittal CT image of the humeroulnar joint. Demonstration of the bone density in intervals of 200 HU. The mechanical stiffness (elastic modulus) of the elements in the FE model was computed from these density values. With permission from Merz B, Eckstein F, Hillebrand S, Putz R (1997) Mechanical implications of humero-ulnar incongruity – finite element analysis and experiment. J Biomech 30:713–721 (Copyright Elsevier Science 1997)

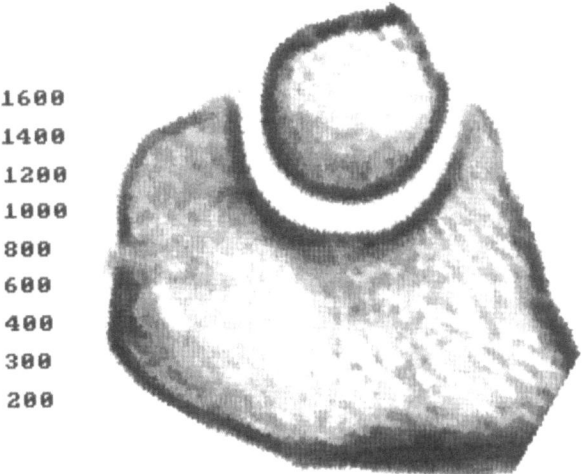

1600
1400
1200
1000
800
600
400
300
200

the sagittal ridge were used for the construction of the two-dimensional FE model (Fig. 11).

A sagittal section was prepared with a saw microtome, following the main ridge of the ulna and the groove of the humerus. The section was digitized, and the cartilage thickness measured by image analyses (Vidas IPS 10, Kontron, Eching, Germany) under high magnification at 20° intervals (accuracy about 0.1 mm) (Fig. 11). Since the

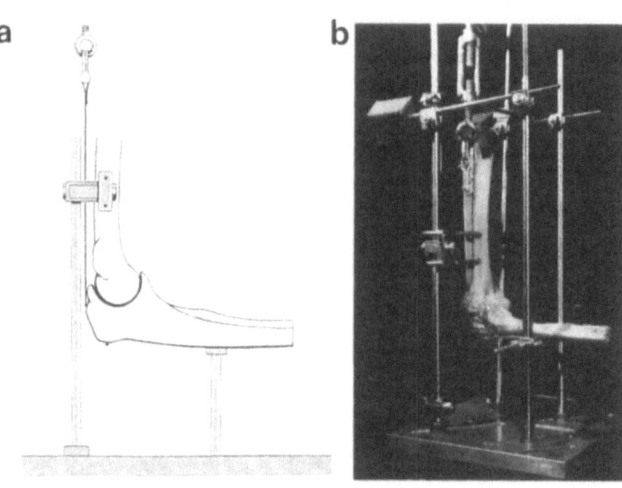

Fig. 10a,b. Design of the biomechanical experiment for the determination of the joint space width, the contact areas and the pressure distribution in the humeroulnar joint. Simulation of extension (musculus triceps brachii) versus resistance at 90° of flexion. **a** Schematic drawing. **b** Photograph

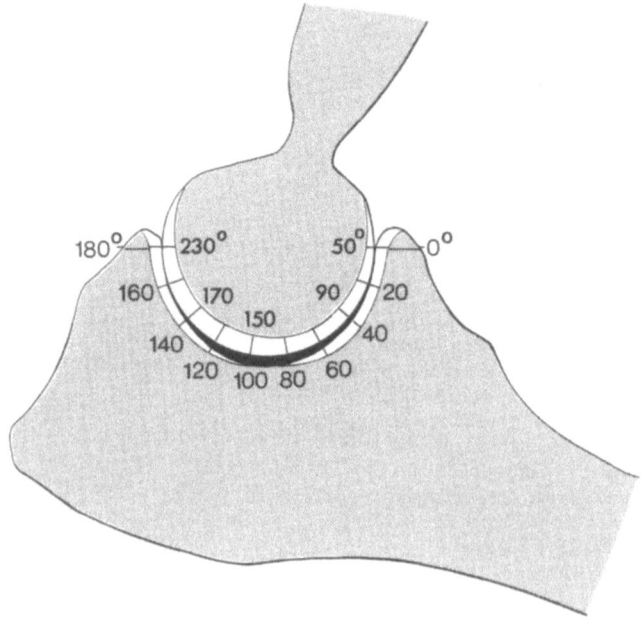

Fig. 11. Coordinate points at which the joint space width (thickness of the polyether cast) and the cartilage thickness was determined for the construction of the anatomically based FE model. With permission from Merz B, Eckstein F, Hillebrand S, Putz R (1997) Mechanical implications of humero-ulnar incongruity – finite element analysis and experiment. J Biomech 30:713–721 (Copyright Elsevier Science 1997)

articular surface of the humerus is circular in a sagittal section (Shiba et al. 1988; Eckstein et al. 1995a), the geometric model was constructed from this surface, the width of the joint space and the cartilage thickness being taken from the above measurements.

From a sagittal CT section we obtained the inhomogeneous distribution of the elastic modulus, which differed for each separate element (Fig. 12a). As with other density-based analyses (Merz et al. 1996), a linear relationship was assumed between

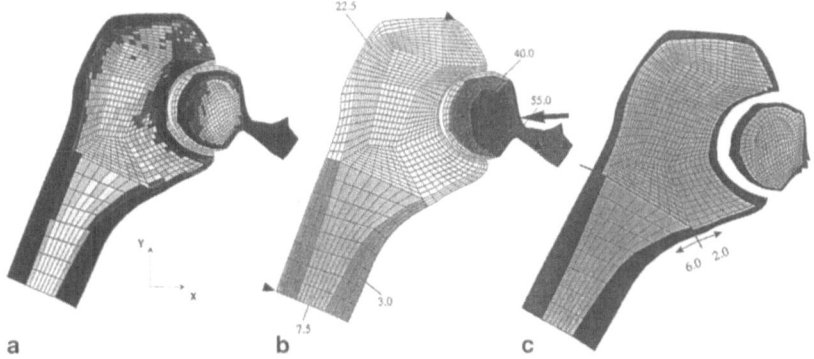

Fig.12a–c. Anatomically based FE model of the humeroulnar joint. **a** "Frontplate" of the model with quantitatively accurate representation of the joint space width (incongruity), cartilage thickness and the inhomogeneous distribution of the bone stiffness (see Figs. 8, 9). **b** Thickness of the frontplate according the dimensions of the anatomical specimen (in mm). **c** "Sideplate" of the model (thickness in mm). The frontplate and sideplate are mechanically coupled at the two outer rows of elements (*dark elements*). With permission from Merz B, Eckstein F, Hillebrand S, Putz R (1997) Mechanical implications of humero-ulnar incongruity – finite element analysis and experiment. J Biomech 30:713–721 (Copyright Elsevier Science 1997)

the Hounsfield values and the density. A quadratic relationship between the density and stiffness was employed for density values of less than 0.4 g/cm^3 (Rice et al. 1988) and a cubic relationship for density values higher than that (Carter and Hayes 1977). A locally variable thickness of the model was selected which depended on the actual thickness of the bone (Figs. 8, 12b). To account for the tubular architecture of the long bones and the connection between the subchondral plate and the cortical bone we used a humeral and ulnar "side plate" (Verdonshot and Huiskes 1990) with a thickness of 6 mm in the shaft region and of 2 mm in the metaphysis and epiphysis (elastic modulus 8000 MPa; Fig. 12c). The side plate was connected to the actual model (the "front plate") by the two peripheral rows of elements.

For the simulation of the actual loading conditions, the humerus was first pressed centrally into the ulna. The nodes in the region of the insertion of the triceps at the olecranon were locked in the x direction to simulate the extension of the arm against resistance during the experiment. The shaft of the ulna was also locked at the distal nodes in a direction perpendicular to its axis. However, to ensure an unrestrained adjustment of the notch and the trochlea, the nodes near the insertion of the triceps and those of the distal end of the shaft were free to move in the y direction of the model (the axis of the shaft). Forces of 125, 250, 375, and 500 N were applied. For the calculations we used the FE program ANSYS 5.1 (Swanson Analysis Systems, Houston, Texas, USA). The nonlinear contact was calculated, using a frictionless two-dimensional contact element. With each of the load magnitudes mentioned above, the pressure distribution at the joint surface (compressive nodal stresses), the maximum compression stresses in the subchondral bone, the maximum tensional stresses and the distribution of the SED in the subchondral bone were calculated in the model.

To analyze the effect of the inhomogeneous material properties on the stresses and strains, a second comparative model was constructed with a homogeneous stiffness of 8000 MPa for the cortical, 500 MPa for the trabecular, and 1800 MPa for the subchon-

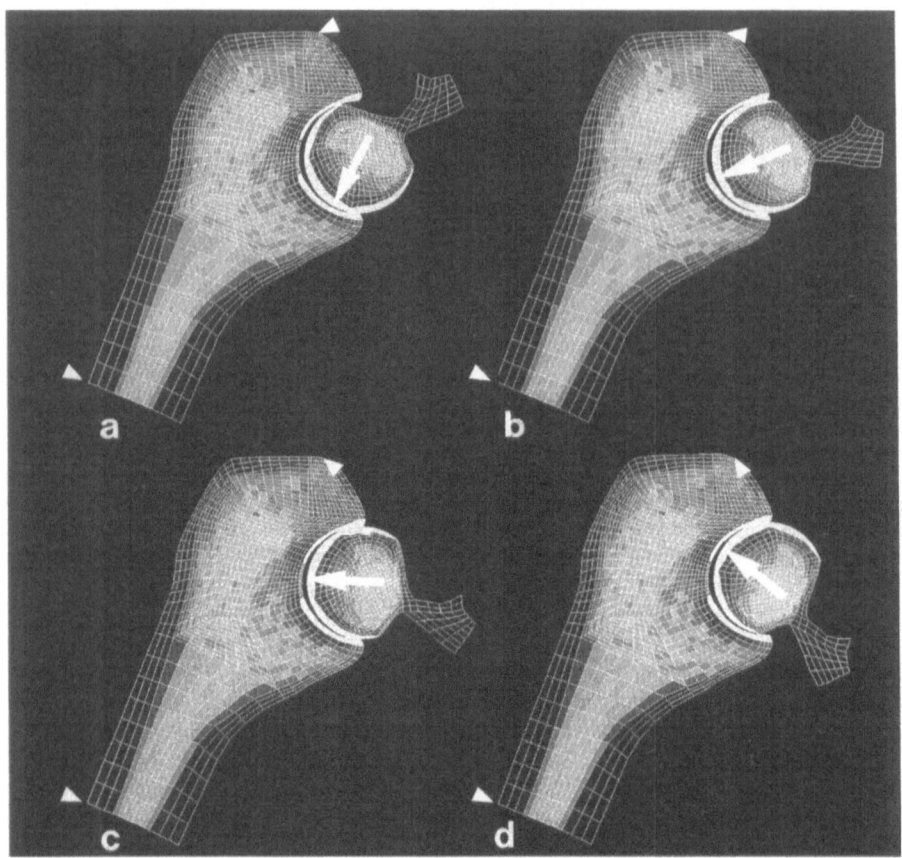

Fig. 13a–d. Simulation of loading and boundary conditions at 30° (**a**), 60° (**b**), 90° (**c**), and 120° (**d**) flexion angles in the anatomically based model of the humeroulnar joint. *Arrows*, direction of loading; *triangles*, pull of the musculus triceps brachii (movement along the base of the triangles is possible, movement in the direction of their tips is locked). The ulnar shaft can move along its long axis, but rotation movements are restricted. *Gray values*, material properties; *dark shading*, high stiffness; *light shading*, low stiffness

dral bone (Choi et al. 1990). The element generation, geometry of the model, and boundary conditions were otherwise identical to those of the inhomogeneous model. The convergence of the model was tested, and the results were compared to those of a model in which each element was replaced by four elements.

In a further step, the loading at flexion angles of 30°, 60°, and 120° were simulated. For this the shaft of the ulna was rotated relative to the humerus, and the force applied in the direction of the olecranon (120° flexion angle), and in the direction of the coronoid process (30° and 60° of flexion) (Fig. 13). The pressure distribution throughout the joint surface, the maximum compressive and tensional stresses in the subchondral bone, and the distribution of the SED were again calculated as described above, as well as the principal stresses throughout the model. The agreement and deviation between these parameters was assessed by computing the correlation coefficient (r) for the values obtained from the proximal to the distal end of the ulnar surface.

3.4
Experimental Validation

After obtaining the polyether casts under a force of 25 N, the same procedure (Fig. 10a, b) was used to prepare further casts from the same specimen at 250, 500, and 750 N, and the contact areas were graphically recorded. Since this provided information about the site of contact, but not about the pressure distribution within the contact areas or the magnitude of the pressure, these were measured with a pressure-sensitive Fuji film (Fukubayashi and Kurosawa 1980; Hehne 1990; Ateshin et al. 1994b). This film consists of two 100-μm layers, one of which contains microspheres of various sizes which are filled with a special chemical. When this layer is pressed against the second, the capsules burst (depending upon the magnitude of this pressure) and stain the second film red. The film used ("ultralow pressure") records contact pressures within the range of 0.5–2.5 MPa. To avoid crinkle artifacts the film was cut into halves in each case, one for the medial and one for the lateral joint surface. Before inserting it into the joint cavity, it was wrapped in very thin polyethylene foil to protect it from moisture. The joint components were then pressed together for 30 s under forces of 250, 500, and 750 N. Since the staining of the film is related to the pressure in a nonlinear manner, a second piece of film was loaded at the same time with a stamp at defined pressures of 0.5, 1.0, 1.5, 2.0, 2.5, and 3.0 MPa. The calibration film and the film from the joint were then digitized with a personal computer (Quadra 700, Apple Macintosh) using a digital camera with a green filter. By means of graphic software (Studio 8) the gray values of the films were then replaced by false colors to visualize 0.5 MPa pressure intervals ranging from less than 0.5 MPa to greater than 2.5 MPa.

3.5
Morphological Investigations on the Elbow Joint

3.5.1
Determination of the Subchondral Mineralization
by Means of CT-OAM

This part of the investigation was undertaken on 36 embalmed dissecting-room specimens of the elbow joint with an age range of 58–97 years (mean 77 years; 19 women, 17 men), specimens with macroscopic evidence of damaged cartilage being excluded from the study. The joints were classified according to the morphology of the joint surface (Tillmann 1971, 1978). Sixteen joints had completely divided surfaces (group A, Fig. 2a), in 12 cases the surface was divided medially but continuous laterally (group B, Fig. 2b) and in 8 joints a continuous surface was present both medially and laterally (group C, Fig. 2c).

Serial sagittal images were obtained from these joints at intervals of 2 mm with a Somatom CT scanner, and the distribution of the subchondral mineralization deter-

Fig. 14a–c. Anatomical definition of the coordinate points for the assembly of a joint template of the human elbow joint. The template allows the average distribution of subchondral bone density in several specimens to be computed (see Figs. 43, 44)

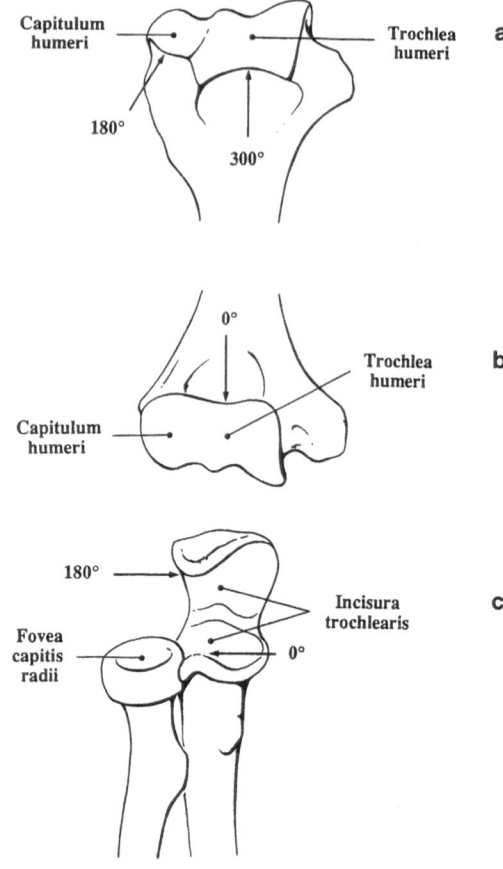

mined by CT-OAM (Müller-Gerbl et al. 1989, 1992; Müller-Gerbl 1998). Eight density regions from less than 300 HU to greater than 900 HU were demonstrated in steps of 100 HU by a so-called "maximal density projection" of the subchondral values on to the bone-cartilage interface. A grade scale (Fig. 14) was used to locate the positions of the density regions on a template of the joint surface. The template made it possible to determine the average distribution pattern for all 36 specimens as well as those for each of the three groups. First a defined gray value was assigned to the density intervals (in each case), and the average gray values at each location were computed by image analyzing software (Adobe Photoshop) at each point of the template. In these summation pictures the original density intervals were restored by reassembling the corresponding gray values.

To test the reproducibility of the method, a single joint surface was scanned six times with CT, the density distribution being reconstructed in each case from the data sets and the distribution patterns compared at 250 coordinate points by image analysis (Vidas, Kontron, Eching, Germany).

3.5.2
Assessment of the Trabecular Architecture

A contact radiograph was prepared on Strukturix mammography film from the midsagittal sections of the joint from which the model had been constructed. Eight more ulnas with subdivided articular surfaces (Fig. 2) were embedded in methyl methacrylate and 500-μm sagittal sections cut at 2-mm intervals with a Leitz microtome. Contact radiographs were made in the manner described above.

3.5.3
Analysis of the Subchondral Split Lines

The split lines can be used to provide information about the preferential orientation of the collagen fibrils. So far this technique has been employed for analyzing the organization of the fibrils in the tangential layer of the articular cartilage (Hultkrantz 1898; Pauwels 1959, 1965, 1980; Molzberger 1973; Tillmann 1978; Akizuki et al. 1986; Jeffery et al. 1991). To be able to evaluate the preferential orientation of the split lines in the subchondral bone, the cartilaginous layer must first be removed and the bone decalcified. Twenty specimens of the elbow joint taken from the dissecting room (ages 57–94 years) with transversely subdivided ulnar joint surfaces (type A; Tillmann 1971, 1978) were macerated in 5% sodium bicarbonate solution and freed from soft tissue. The specimens were then decalcified in 5% nitric acid, soaked in 5% sodium sulfate, and stored in 96% alcohol.

Finally, the subchondral plate was pierced with pins at right angles at regular intervals, and the split line shown up by polishing with india ink. The direction of the split lines was recorded by photographing them with black-and-white films.

4 Results

4.1
Idealized Model of Concave Joint Incongruity

In the static analysis of the model with a 10% deeper socket, a bicentric pressure distribution was observed with maximal contact stresses at the periphery. The pressure distribution was relatively independent of the direction of the applied force (Figs. 16). With central load application the maxima at the peripheral parts of the joint surface amounted to about 0.9 MPa, whereas the central region between the two contact areas encountered no pressure (Figs. 15a, 16). With oblique load application (22.5° and 45°) the pressure distribution was also bicentric, and it was interesting to note that in this case the load bearing areas were located at exactly the same place as with the central load application (Fig. 16). The size of the contact pressure varied

Fig. 15a,b. Static analysis of load transmission in the idealized model with a 10% deeper joint socket (70 N, central load application).
a Vector plot of the first principal stress (mainly compressive).
b Vector plot of the second principal stress (mainly tensile)

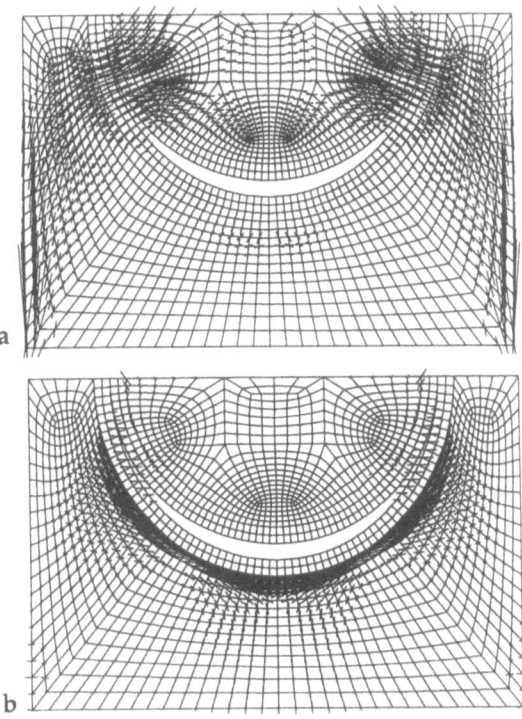

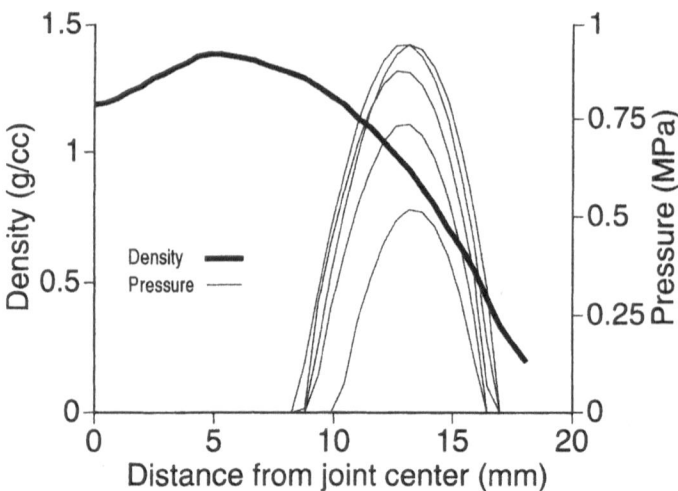

Fig. 16. Compressive stress (*thin lines*) at the articular surface of the idealized model with a 10% deeper joint socket for five different load cases (see Fig. 6) and subchondral bone density (*thick line*) in the same model after simulation of 300 days of bone remodeling. *Y-axis*, density (g/cc) and the contact pressure (MPa); *x-axis*, distance from the center of the joint surface (0 mm=center, 20 mm=margin). Note that only one half of the socket is shown, and that the other half is symmetric. With permission from Jacobs CR, Eckstein F (1997) Computer simulation of subchondral bone adaptation to mechanical loading in an incongruous joint. Anat Rec 249:317–326 (Copyright Wiley-Liss, Inc., a division of John Wiley & Sons Inc., 1997)

between 0.5 and 1.0 MPa, depending upon whether the applied force was orientated towards the load bearing area or to the other side of the joint.

Figure 15 shows, in addition to the pressure on the joint surface, the size and orientation of the principal stresses in the bone of the socket and the head, which gives an impression of the flow of force within it. The vectors which are orientated perpendicular to the joint surface (within an area about 8–17 mm lateral to the center of the socket) represent the compressive stresses at the articular surface (see also Fig. 17). However, the vectors in the socket which run tangential to the joint surface (Fig. 15b) show a significantly higher value, particularly within a region 2–8 mm lateral to the center (Figs. 15b, 17). These represent tensional stresses resulting from the socket being spread apart during load transfer due to the incongruity of the joint components.

A bone remodeling simulation over ten time intervals of 30 days [3000 daily cycles each of 70 N, with widely varying loads (all five load cases) and making use of a "dead zone"], leads to a bicentric density distribution in the joint socket (Fig. 18). The maxima amounted to about 1.5 g/cm³ and were located about 6 mm lateral to the center of the joint surface. In the depth of the socket the subchondral density was only about 1.2 g/cm³, in the periphery it was reduced to minimal values (Fig. 20). The distribution of the subchondral density showed high agreement with that of the tangential (tensional) stresses, but not with the normal (compressive) stresses in the subchondral bone, or the pressure distribution over the articular surface (see Figs. 16, 17).

After convergence of the bone remodeling simulation (300 days), the density distribution depended only slightly upon the initially chosen distribution of the bone

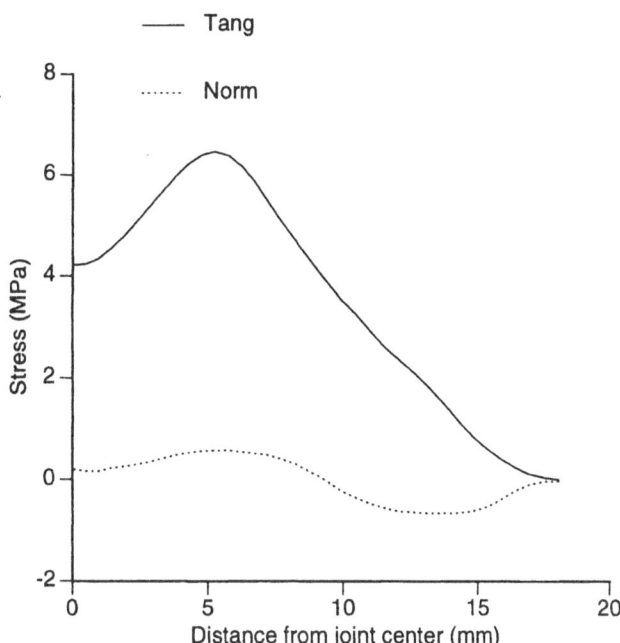

Fig. 17. Magnitude of the normal stress in the subchondral bone (perpendicular to the bone cartilage interface) and of the tangential stress (parallel to the bone cartilage interface) as a function of the distance of the center of the joint socket (*positive values*, tension; *negative values*, compression). The distribution of the tangential stresses (mainly tensile), but not that of the normal stress (mainly compressive) resembles that of subchondral bone density (for comparison see Fig. 16). With permission from Jacobs CR, Eckstein F (1997) Computer simulation of subchondral bone adaptation to mechanical loading in an incongruous joint. Anat Rec 249:317–326 (Copyright Wiley-Liss, Inc., a division of John Wiley & Sons Inc., 1997)

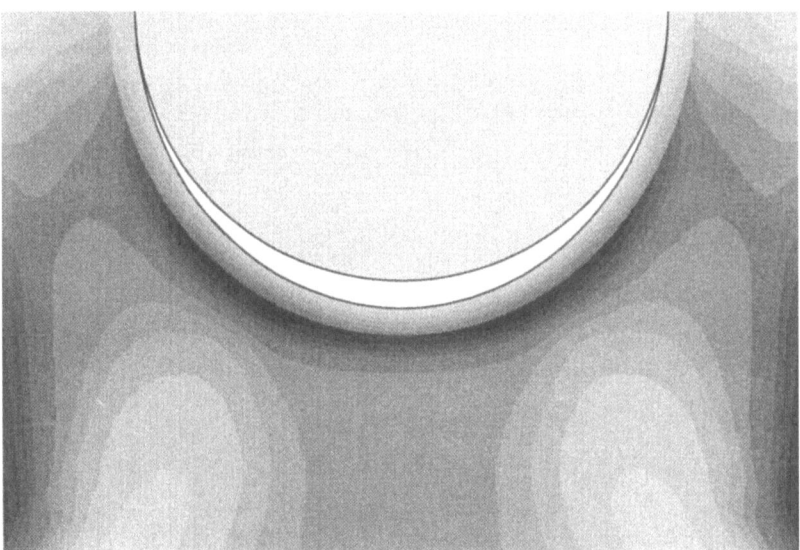

Fig. 18. Result of a computer simulation of 300 days of subchondral bone remodeling in the idealized "concavely" incongruous model with a 10% deeper joint socket (see Fig. 6). *Dark elements*, dense bone (max. 2.0 g/cm³); *light elements*, the opposite (min. 0.2 g/cm³). The load magnitude was 70 N (five load cases as shown in Fig. 6), the cartilage stiffness 15 MPa, the initial density 0.5 g/cm³ for all bone elements and the "dead zone" 20% of the reference value. (These parameters are listed in subsequent figures only when they differ from these). With permission from Jacobs CR, Eckstein F (1997) Computer simulation of subchondral bone adaptation to mechanical loading in an incongruous joint. Anat Rec 249:317–326 (Copyright Wiley-Liss, Inc., a division of John Wiley & Sons Inc., 1997)

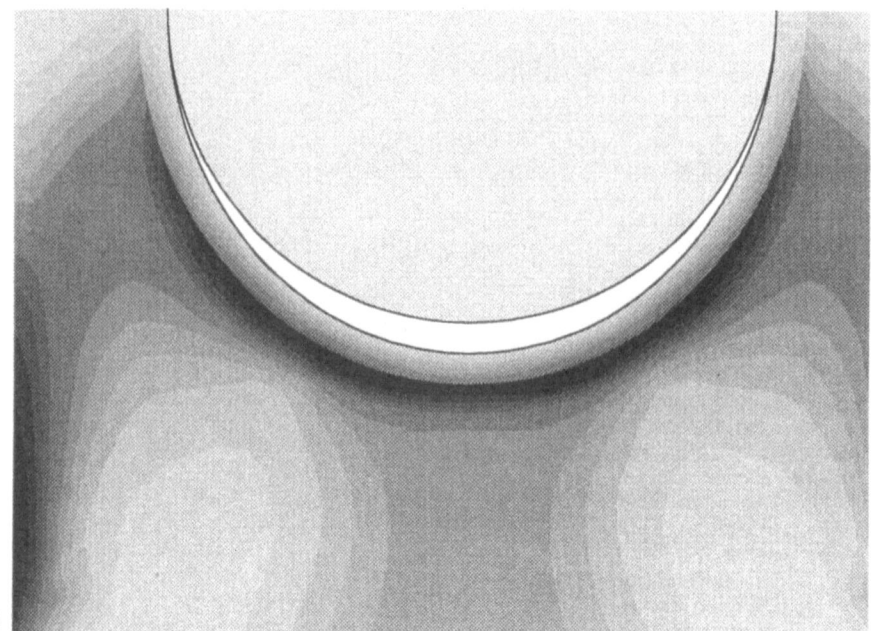

a

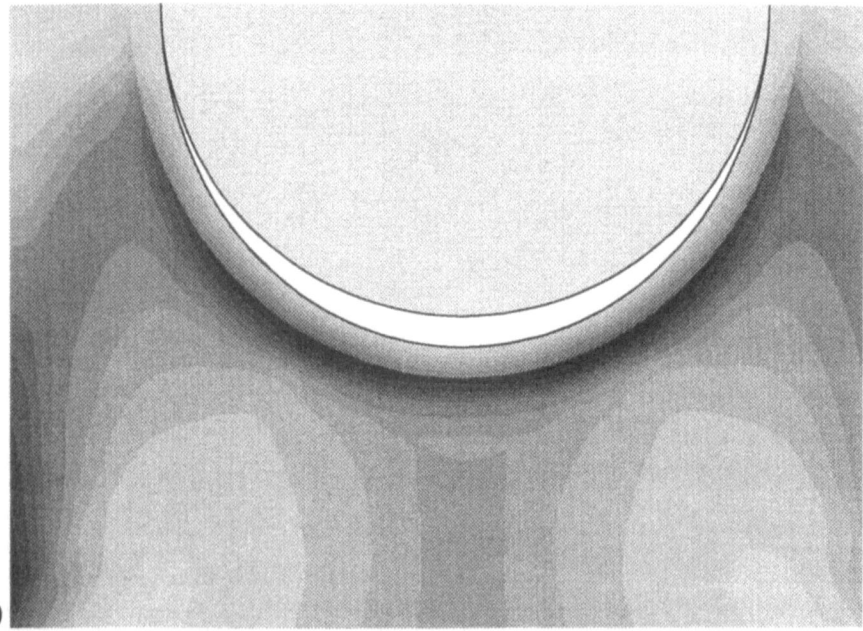

b

Fig. 19a–d. Uniqueness of the bone remodeling simulations (dependence on the initial density distribution) with the use of a "dead" zone (for explanation of "dead" zone see Fig. 5). **a** Initial density 0.5 g/cm^3. **b** Initial density 0.2 g/cm^3. With permission from Jacobs CR, Eckstein F (1997) Computer simulation of subchondral bone adaptation to mechanical loading in an incongruous joint. Anat Rec 249:317–326 (Copyright Wiley-Liss, Inc., a division of John Wiley & Sons Inc., 1997)

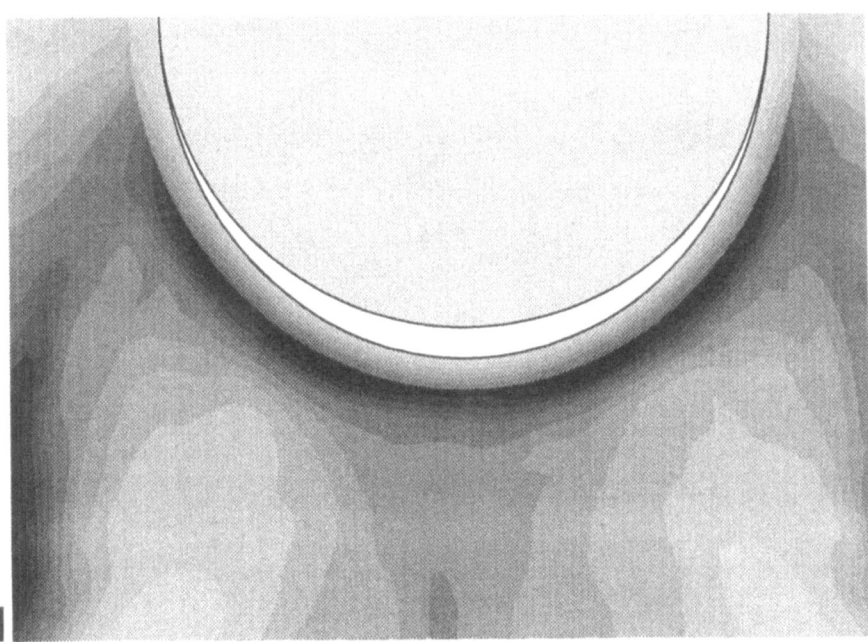

Fig. 19 c,d. c Initial density 2.0 g/cm³. **d** Initial density random for each element (between 0.2 and 2.0 g/cm³). With permission from Jacobs CR, Eckstein F (1997) Computer simulation of subchondral bone adaptation to mechanical loading in an incongruous joint. Anat Rec 249:317–326 (Copyright Wiley-Liss, Inc., a division of John Wiley & Sons Inc., 1997)

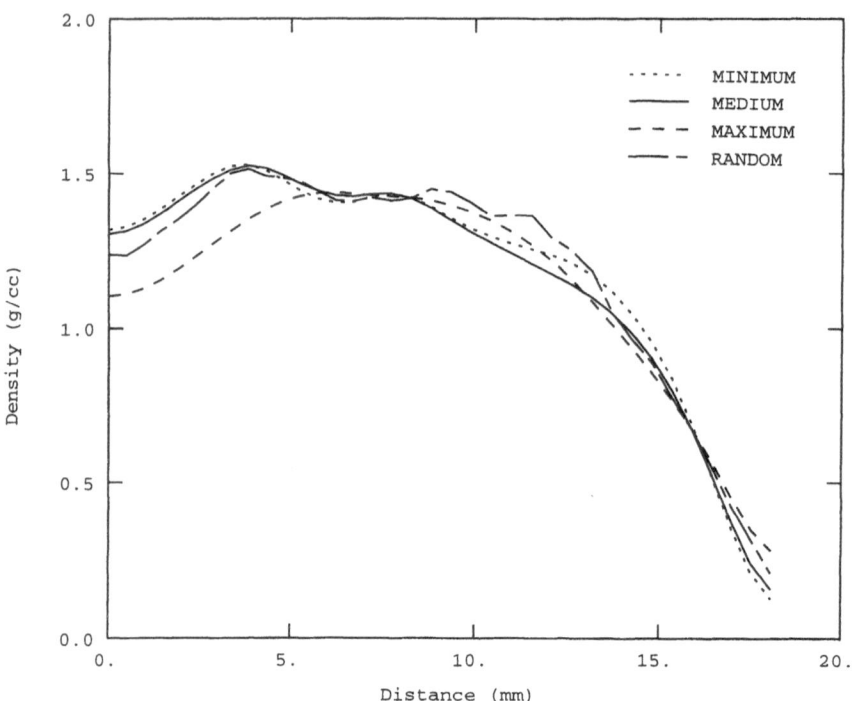

a

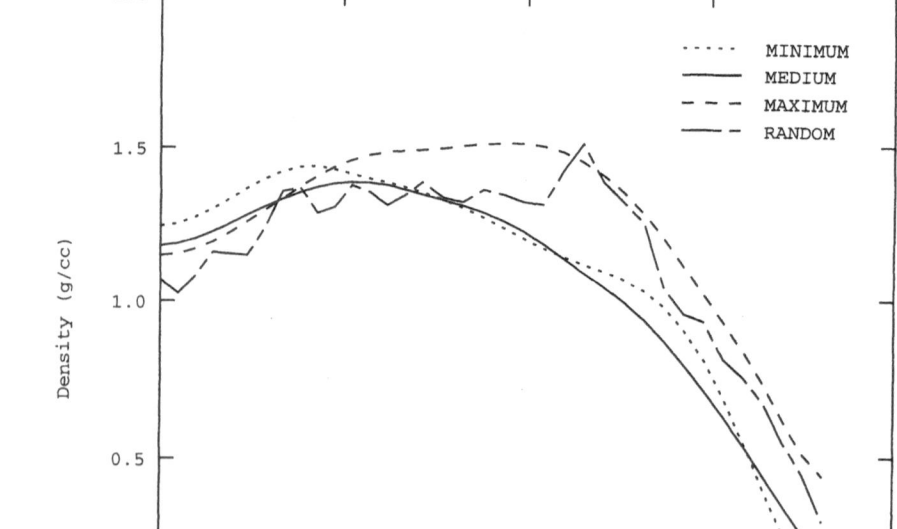

b

density in the socket. Figure 19 shows the density patterns without the use of a "dead zone" at the start of the models of medium density (0.5 g/cm³; Fig. 19a), minimal density (0.2 g/cm³; Fig. 19b), maximal density (2.0 g/cm³; Fig. 19c) and each element of a randomly distributed density between 0.2 and 2.0 g/cm³ (Fig. 19d). If no "dead zone" was included, the resulting density distribution patterns within the subchondral bone were nearly identical (Fig. 20a). With a 20% "dead zone" (Fig. 20b) the density patterns were less similar, but the typical bicentric distribution was reproduced in every case.

The stiffness of the articular cartilage also had no significant effect on the density. The subchondral density values obtained with stiffer cartilage (25 MPa) were marginally higher (+0.05 g/cm³) and with softer cartilage (5 MPa) somewhat lower (–0.2 g/cm³); however, the relative distribution of the values under the articular surface was almost identical (Fig. 21).

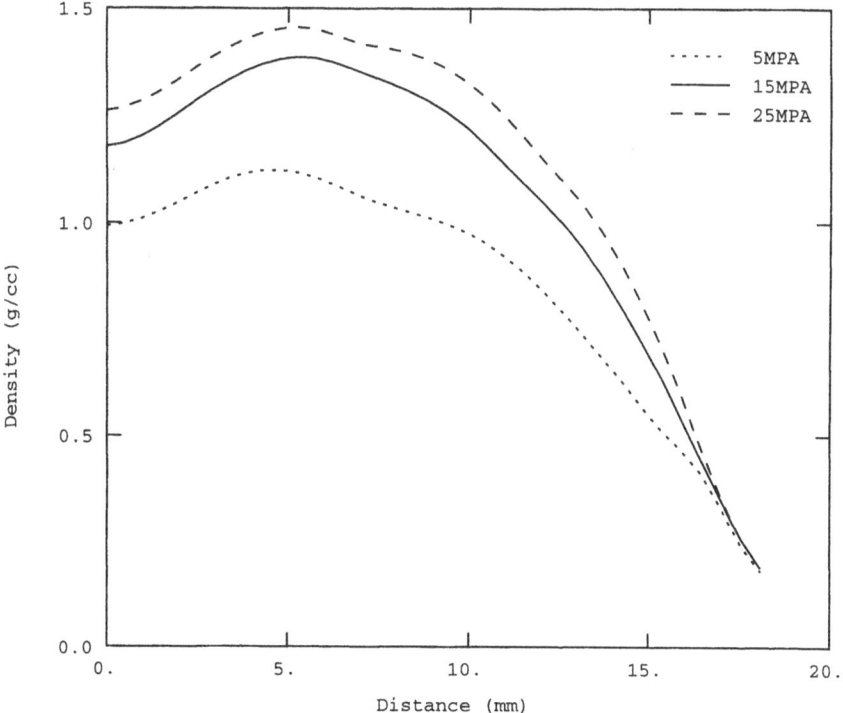

Fig. 21. Dependence of the simulation on the stiffness of the articular cartilage (5, 15, and 25 MPa). With permission from Jacobs and Eckstein 1997 (see above)

Fig. 20. Uniqueness of the bone remodeling simulations without (**a**) and with (**b**) the use of a "dead" zone (for explanation of "dead" zone see Fig. 5). The magnitude of the subchondral bone density is plotted as a function of the distance from the joint center (center, 0 mm; joint margin, 18 mm). Note that only one half of the socket is shown and that the other half is symmetric (for comparison also see Fig. 19). With permission from Jacobs CR, Eckstein F (1997) Computer simulation of subchondral bone adaptation to mechanical loading in an incongruous joint. Anat Rec 249:317–326 (Copyright Wiley-Liss, Inc., a division of John Wiley & Sons Inc., 1997)

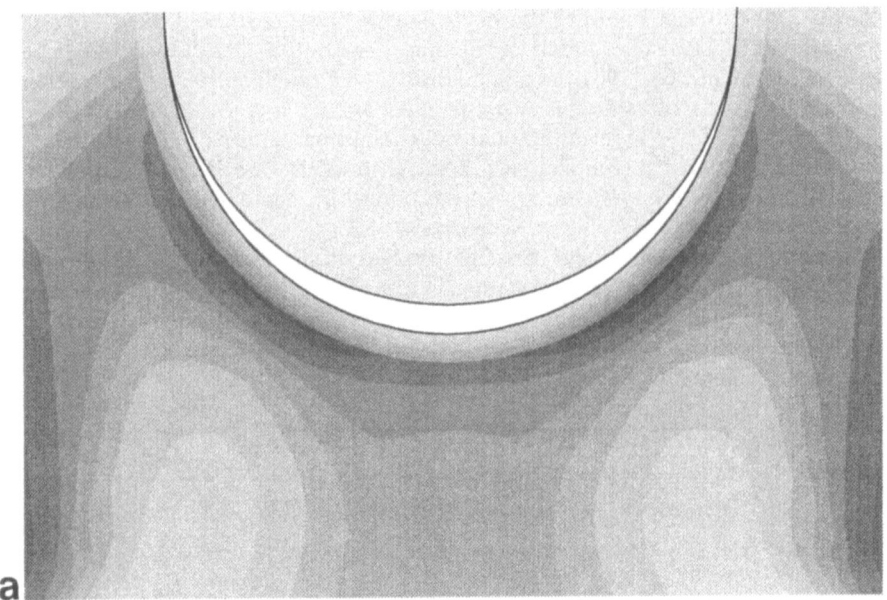

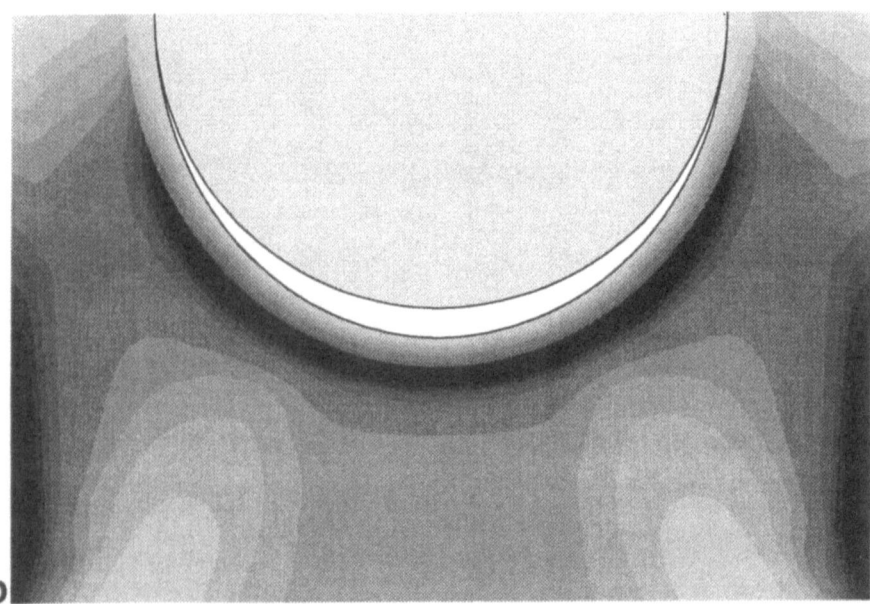

Fig. 22. Distribution of density in the joint socket of idealized model with a 10% deeper socket at loads of 35 N (a) and 140 N (b)

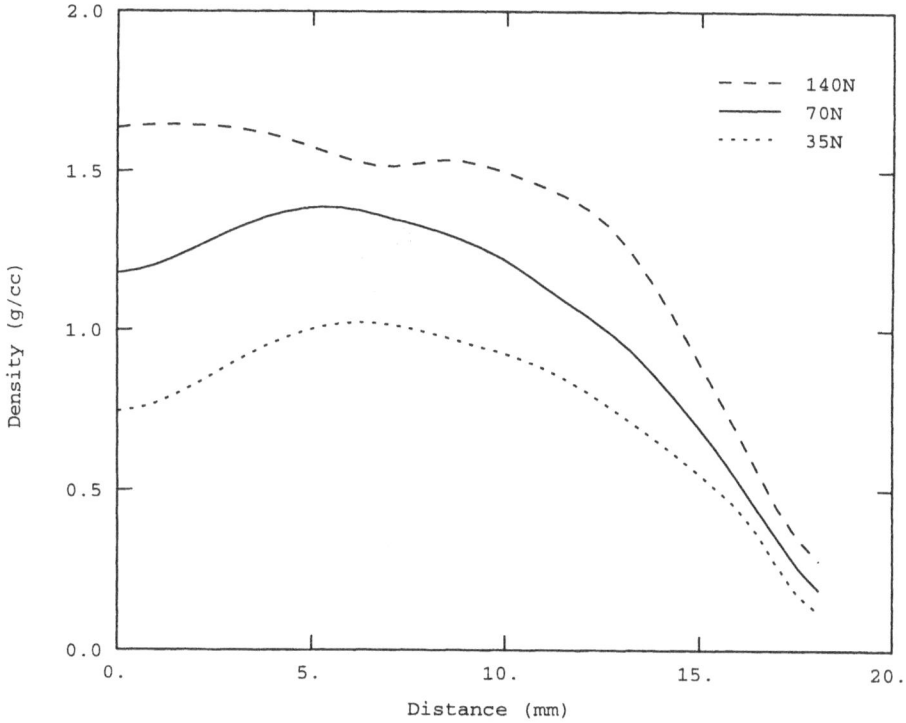

Fig. 23. Dependence of the simulation on the magnitude of loading (35, 70, and 140 N)

With a reduction in the load to 35 N (Fig. 22), the absolute value of the subchondral density was lower, although the bicentric density pattern remained unaltered (Fig. 23). With a load of 140 N, however, not only was there a greater density, but the distribution pattern was also changed. The highest values were now located in the center and fell off in a bell-shaped curve towards the periphery (Figs. 22b, 23).

With 140 and with 70 N no contact of the head was recorded at the center of the socket; the static analysis nevertheless revealed that the central tangential stresses assumed significantly higher values than with 70 N (Figs. 17, 24), since the socket was more markedly spread apart. Lowering the number of loading cycles to 1000 per day had a similar effect to reducing the load to 35 N, and raising it to 10,000 cycles had a comparable although less marked effect than raising the load to 140 N (Fig. 25).

Fig. 24a,b. Static analysis of load transmission in the idealized model with a 10% deeper joint socket and with a load of 140 N (central load application). **a** Vector plot of the first principal stress (mainly compressive). **b** Vector plot of the second principal stress (mainly tensile). For comparison see Fig. 15

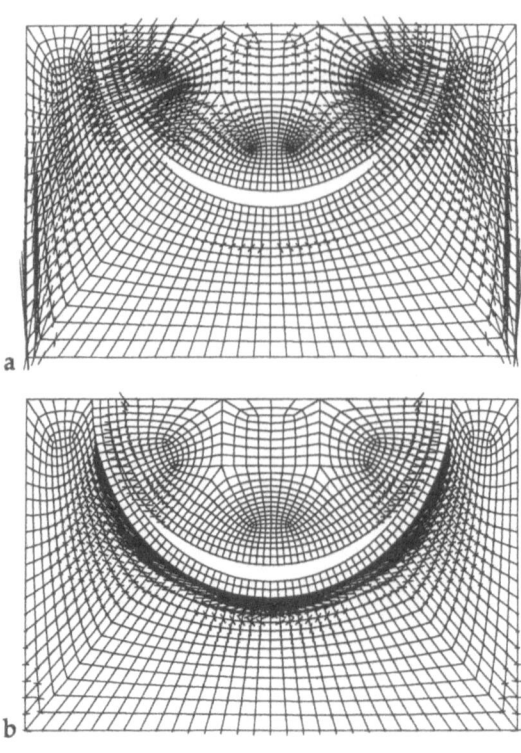

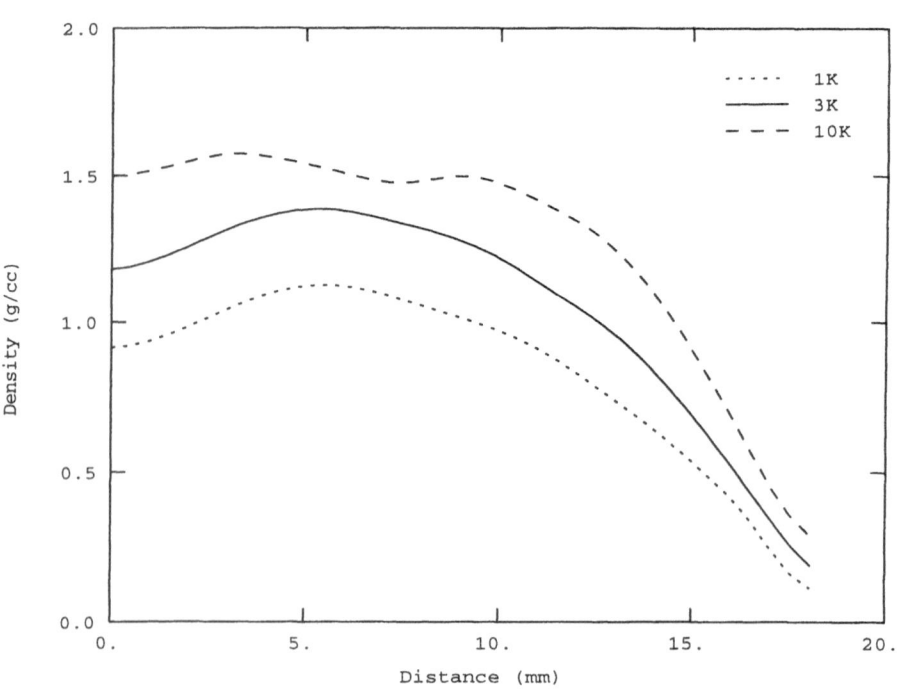

4.2
Idealized Comparative Geometric Models

In comparison with the concavely incongruous configuration, the analysis of the principal stresses in the congruous joint socket revealed a similar magnitude of the compressive stresses, but significantly lower (tensional) stresses (Fig. 26). In the bone remodeling simulation, significant differences between the models were seen when all five load cases were employed (70 N, 3000 daily cycles, initial density 0.5 g/cm³, cartilage stiffness 15 MPa, "dead zone" 20%) (Figs. 18, 27). Simulations in models with deeper joint sockets (Fig. 27c, d) always produced a bicentric density distribution of the subchondral bone, the absolute density taking on higher values with increasing concave incongruity (Fig. 28). In the congruous case and with convex incongruity (wider sockets) the subchondral density values were lower and more homogeneously distributed (Figs. 27a, b, 28).

With the introduction of a single (central) load the differences between the models became more pronounced. In the congruous model the density fell off slowly, and in

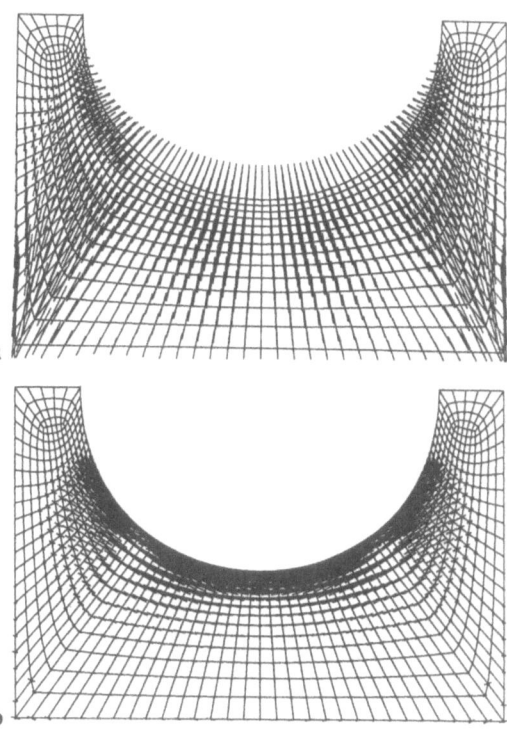

Fig. 26a,b. Static analysis of load transmission in the idealized model with a congruous geometric configuration (70 N, central load application, head not shown). **a** Vector plot of the first principal stress (mainly compressive). **b** Vector plot of the second principal stress (mainly tensile). For comparison see Fig. 15

Fig. 25. Dependence of the simulation on the number of daily load cycles (1000, 3000, and 10,000 cycles). With permission from Jacobs CR, Eckstein F (1997) Computer simulation of subchondral bone adaptation to mechanical loading in an incongruous joint. Anat Rec 249:317–326 (Copyright Wiley-Liss, Inc., a division of John Wiley & Sons Inc., 1997)

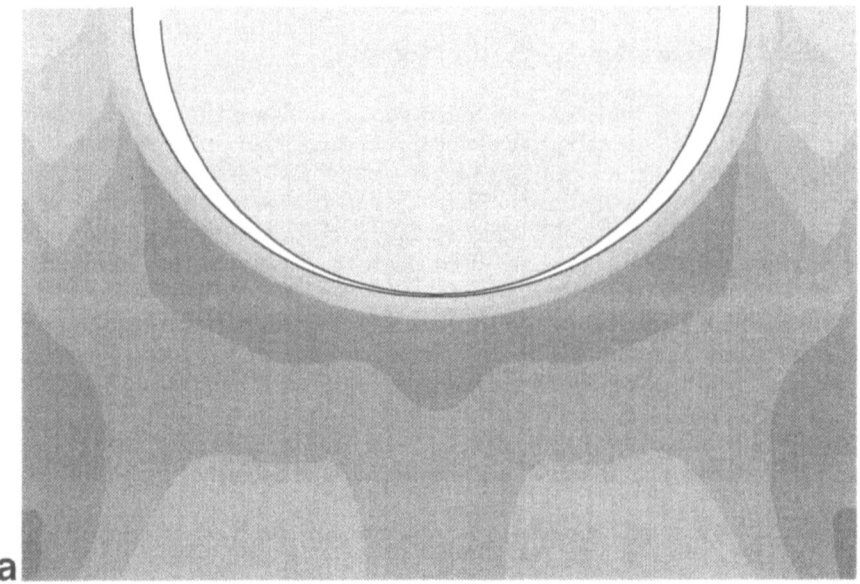

Fig. 27a–d. Distribution of density in the joint sockets of the comparative geometric models (Fig. 7). The load magnitude was 70 N (five load cases according to Fig. 6), the cartilage stiffness 15 MPa, the initial density 0.5 g/cm^3 for all bone elements, and the "dead zone" 20% of the reference value. (For comparison with the reference model, see Fig. 18). **a** "Convex" incongruity (10%). **b** Congruity

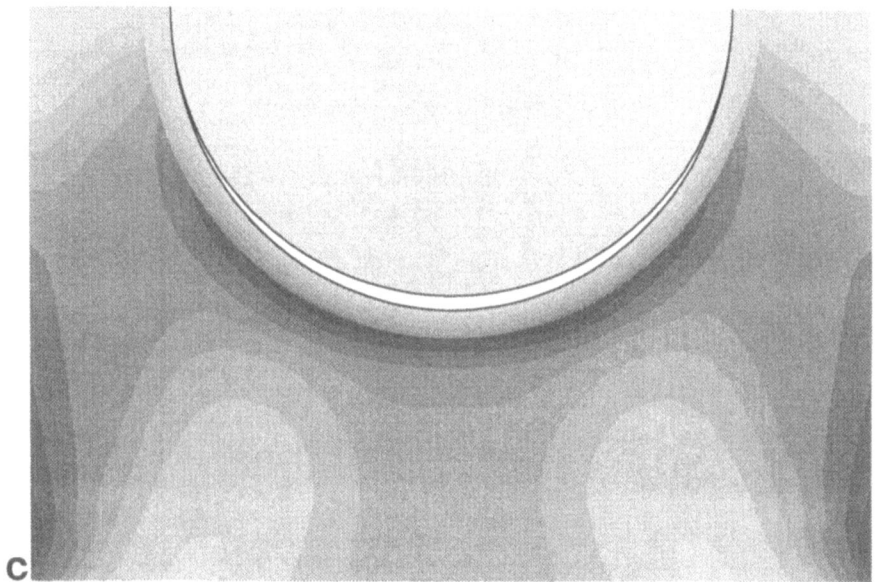

c

d

Fig. 27c,d. Distribution of density in the joint sockets of the comparative geometric models (Fig. 7). The load magnitude was 70 N (five load cases according to Fig. 6), the cartilage stiffness 15 MPa, the initial density 0.5 g/cm3 for all bone elements, and the "dead zone" 20% of the reference value. (For comparison with the reference model, see Fig. 18). **c** "Concave" incongruity (5%). **d** "Concave" incongruity (20%)

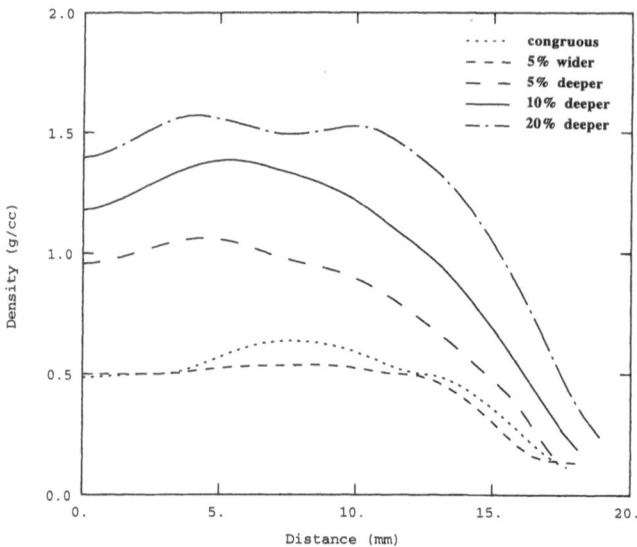

Fig. 28. Widespread loading. Dependence of the simulation on the geometric configuration of the joint [congruity, convex incongruity (wider socket), 5%, 10%, and 20% concave incongruity (deeper socket)]. With permission from Eckstein F, Jacobs CR, Merz BR (1997) Mechanobiological adaptation of subchondral bone as a function of joint incongruity and loading. Med Eng Phys 19:720–728 (Copyright Elsevier Science 1997)

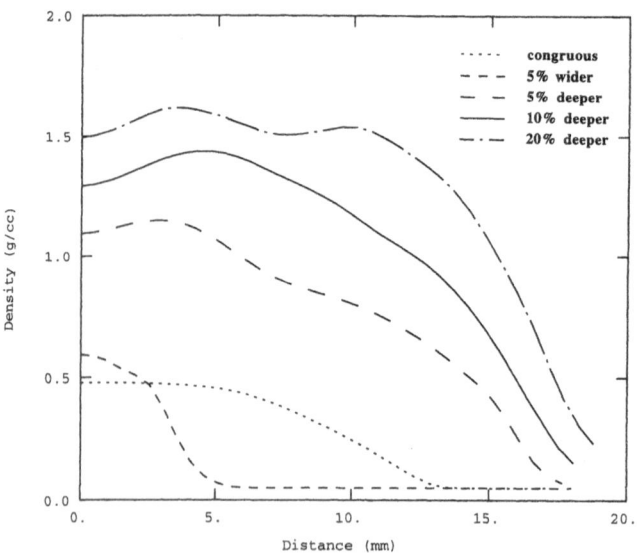

Fig. 29. Central loading. Dependence of the simulation on the geometric configuration of the joint [congruity, convex incongruity (wider socket), 5%, 10%, and 20% concave incongruity (deeper socket)]. With permission from Eckstein F, Jacobs CR, Merz BR (1997) Mechanobiological adaptation of subchondral bone as a function of joint incongruity and loading. Med Eng Phys 19:720–728 (Copyright Elsevier Science 1997)

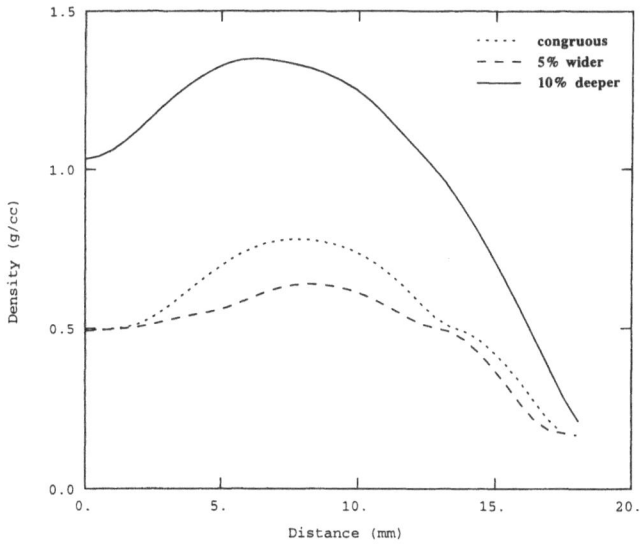

Fig. 30. Bicentric loading. Dependence of the simulation on the geometric configuration of the joint [congruity, 10% convex incongruity (wider socket), and 10% concave incongruity (deeper socket)]. With permission from Eckstein F, Jacobs CR, Merz BR (1997) Mechanobiological adaptation of subchondral bone as a function of joint incongruity and loading. Med Eng Phys 19:720–728 (Copyright Elsevier Science 1997)

the convexly incongruous model rapidly, from the center to the periphery (Fig. 29). With concave incongruity, on the other hand, a slight bicentric density distribution was observed in all cases; even with a purely central loading. It is also remarkable that in these models the density at the periphery of the joint remained much higher than with congruity or with convex incongruity (Fig. 29).

With bicentric load application (load cases 3a and 3b, 45°) a bicentric density distribution was seen in all models (Fig. 30), even with congruity and convex incongruity. With asymmetrical loading (load cases 1, 2a, 3a) significant differences between the models were again observed. With wider and congruent sockets the density assumed higher values on the side of the model where the load was applied and almost vanished on the other side (Fig. 31a). With concave incongruity, on the other hand, the subchondral density maintained a much greater value on the unloaded side (Fig. 31b).

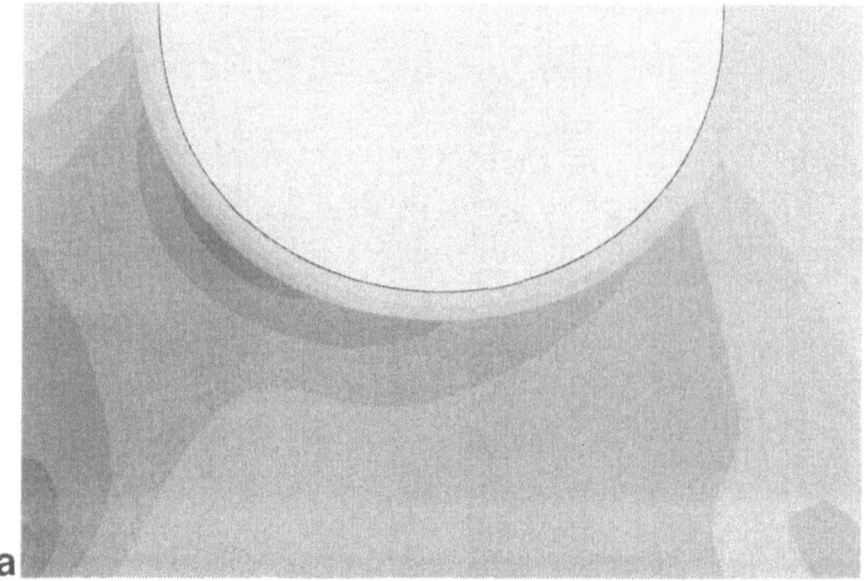

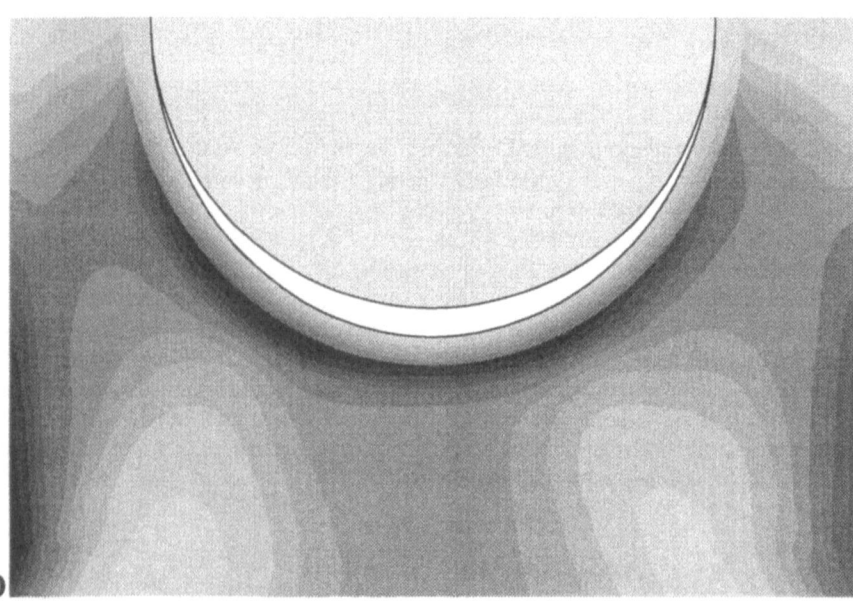

Fig. 31a,b. Asymmetric loading. Dependence of the simulation on the geometric configuration of the joint. **a** Congruity. **b** 10% concave incongruity. With permission from Eckstein F, Jacobs CR, Merz BR (1997) Mechanobiological adaptation of subchondral bone as a function of joint incongruity and loading. Med Eng Phys 19:720–728 (Copyright Elsevier Science 1997)

4.3
Anatomically Based Model of the Humeroulnar Joint

With central loading of 125 N at 90° flexion, the anatomical model of the humeroulnar joint made contact in the ventral and dorsal regions of the articular surface (Fig. 32a). With a greater load, the contact areas spread out towards the depth of the trochlear notch. Whereas a small joint space could still be observed under 375 N, at 500 N the head and socket made complete contact (Figs. 32c, d). At this load the pressure in the dorsal part of the joint surface; at the olecranon amounted to 3.6 MPa, and in the ventral part, at the coronoid process, to 2.4 MPa. Although a greater contact pressure appeared dorsally, the size of the ventral and dorsal load-bearing areas (>0.5 MPa) was roughly the same. In the subchondral bone, maximal compressive stresses from 1.3 MPa (125 N) to 3.1 MPa (500 N) were recorded, also with a bicentric distribution showing maxima in the ventral and dorsal regions of the joint (Fig. 32). The values in

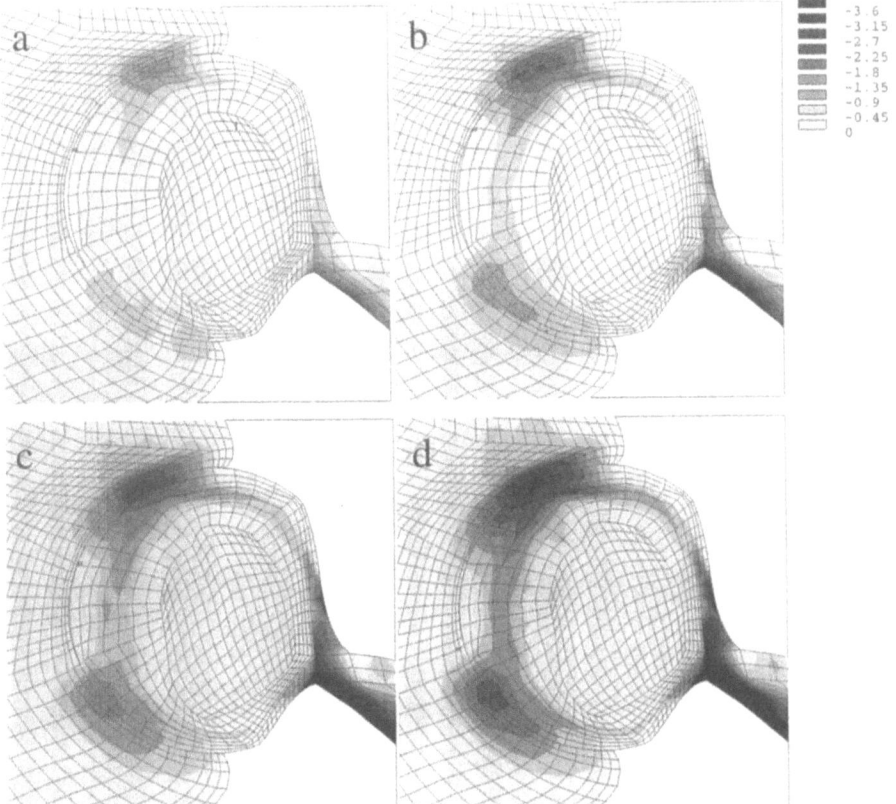

Fig. 32. Compressive stresses in the anatomically based model of the humeroulnar joint at loads of 125 N (**a**), 250 N (**b**), 375 N (**c**), and 500 N (**d**) at a 90° flexion angle. There is a bicentric (ventrodorsal) distribution of contact stress with minimal pressure in the center (depth) of the trochlear notch. With permission from Merz B, Eckstein F, Hillebrand S, Putz R (1997) Mechanical implications of humero-ulnar incongruity – finite element analysis and experiment. J Biomech 30:713–721 (Copyright Elsevier Science 1997)

the model with a homogeneous density distribution within the cortical, trabecular, and subchondral bone were very similar to those of the inhomogeneous model.

Considerable tensional stress was found in the subchondral bone of the ulna, but not in that of the humerus (Fig. 33). Under smaller forces their magnitude was comparable to that of the compressive stress, but with more severe loading they were somewhat lower. Under 125 and 250 N a slight bicentric distribution of the tensional stress was observed; the maxima, however, were located closer to the center of the trochlear notch than those of the compressive stresses (Figs. 33a, b). With 375 and 500 N there was a wide central banana-shaped maximum (Figs. 33c, d). In the model with homogeneous bone properties a wide central maximum was observed with all loads.

The distribution of SED, which is regarded as a possible stimulus for bone remodeling, was relatively uniform in the subchondral bone of the inhomogeneous case. In the homogeneous case, there was a bicentric distribution with a ventral and a dorsal

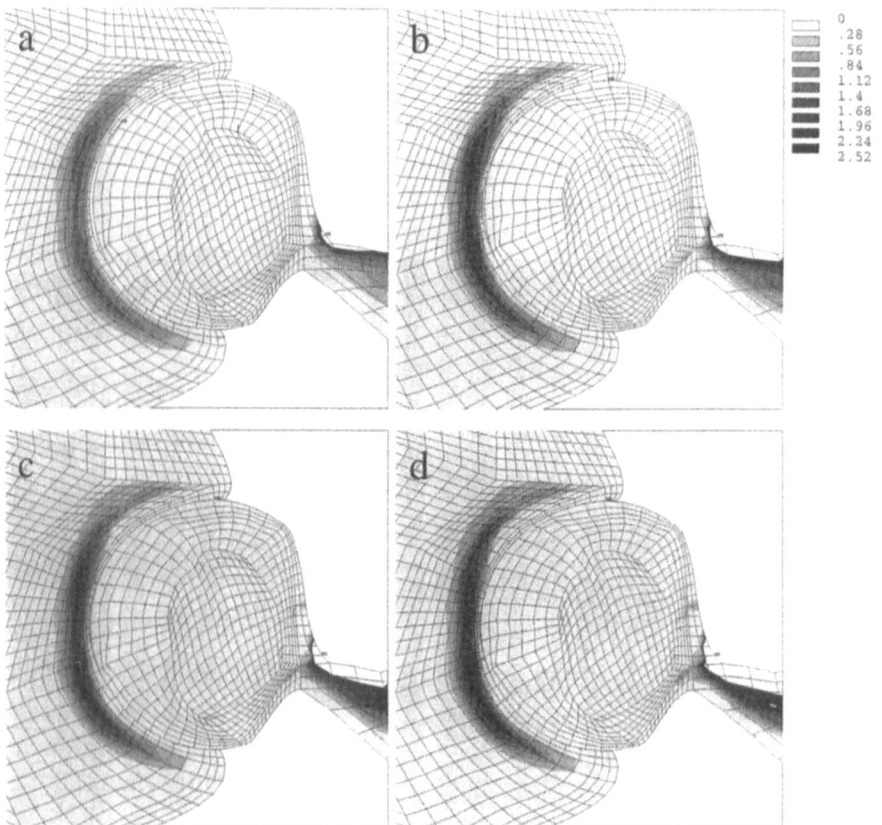

Fig. 33. Tensional stresses in the anatomically based model of the humeroulnar joint at loads of 125 N (a), 250 N (b), 375 N (c), and 500 N (d) at a 90° flexion angle. There are considerable tensional stresses in the trochlear notch, but not in the humerus. With permission from Merz B, Eckstein F, Hillebrand S, Putz R (1997) Mechanical implications of humero-ulnar incongruity – finite element analysis and experiment. J Biomech 30:713–721 (Copyright Elsevier Science 1997)

maximum under 500 N, with significantly higher values being calculated in the ulna than in the humerus (Fig. 34).

With 30°, 60°, and 120° of flexion a bicentric distribution of the contact pressure was observed in all cases (Fig. 35). Corresponding to the course of the joint reaction force, the pressure on the ventral contact area was higher, and on the dorsal surface lower, at small angles of flexion (30° and 60°) than at higher flexion angles (90° and 120°) (Fig. 36).

Very high tensional stresses were found particularly at small angles of flexion (Figs. 37, 38). The SED was also higher at small flexion angles, its distribution showing a moderate, positive correlation with the articular surface pressure and compressive stress in the subchondral bone at 90° flexion (Fig. 39, Table 1), but not at 30°. At small flexion angles, however, the SED was positively correlated with the tensional stress (Fig. 39, Table 1).

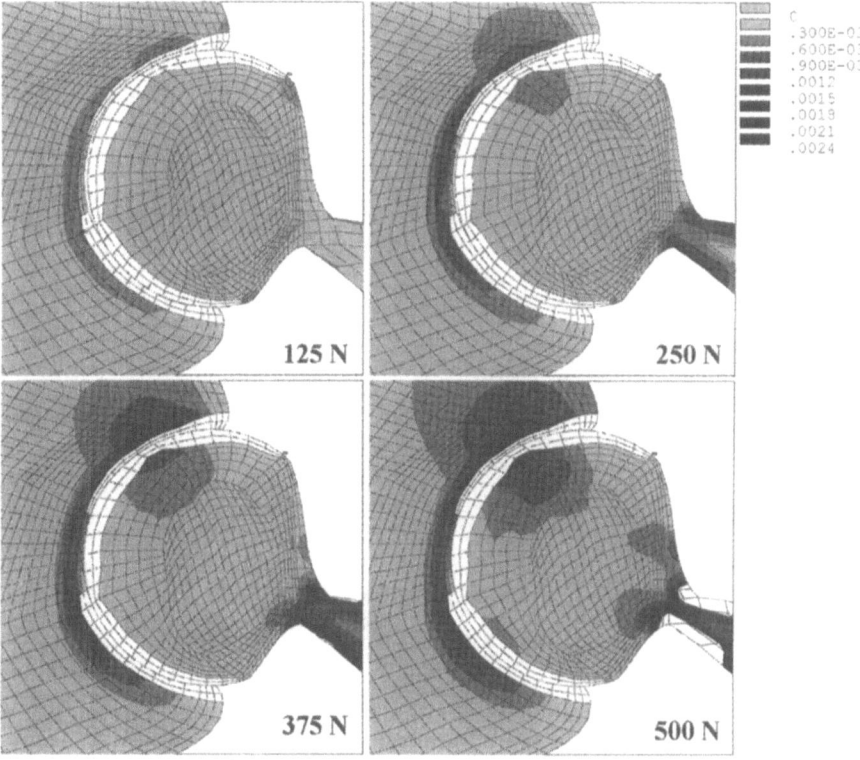

Fig. 34. SED in the anatomically based model of the humeroulnar joint at loads of 125–500 N (90° flexion angle). The cartilage elements are shown, but the SED is only given for the elements representing bone tissue. The plots exhibits ventral and dorsal maxima of SED in the subchondral bone of the trochlear notch. The SED in the notch is considerably higher than that in the humerus. With permission from Merz B, Eckstein F, Hillebrand S, Putz R (1997) Mechanical implications of humeroulnar incongruity – finite element analysis and experiment. J Biomech 30:713–721 (Copyright Elsevier Science 1997)

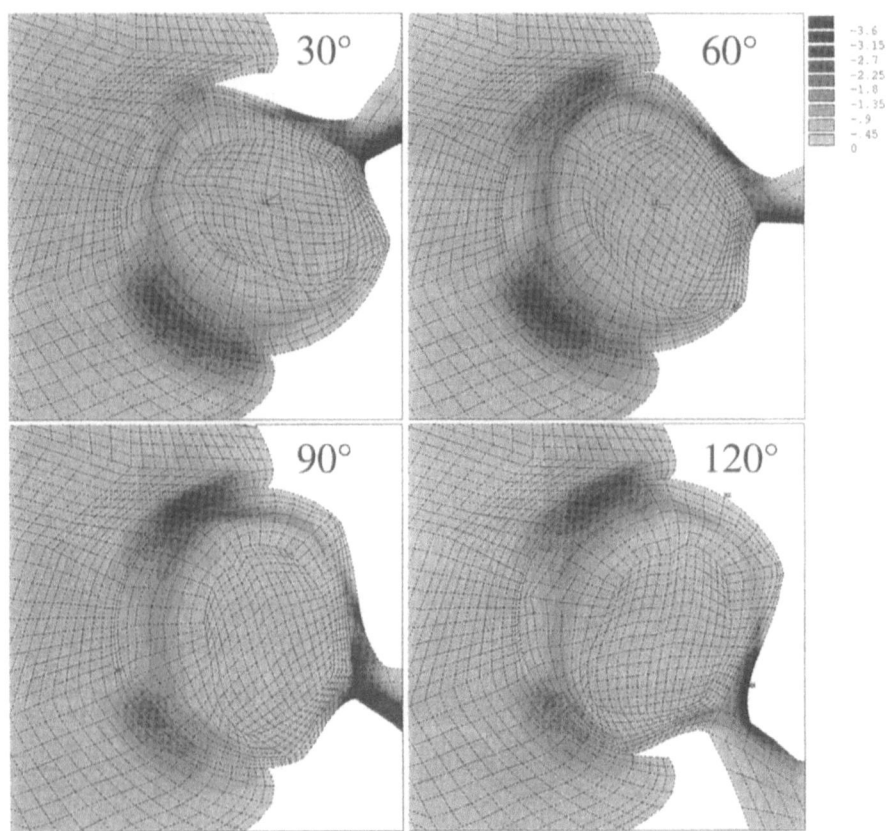

Fig. 35. Compressive stresses in the anatomically based model of the humeroulnar joint at flexion angles of 30°–120° (load=500 N). There is a bicentric (ventrodorsal) distribution of contact stress at all flexion angles. With permission from Eckstein F, Merz B, Schön M, Jacobs CR, Putz R (1999) Tension and bending, but not compression alone determine the functional adaptation of subchondral bone in incongruous joints. Anat Embryol 199:85–97 (Copyright Springer 1999)

Fig. 37. Tensional stresses in the anatomically based model of the humeroulnar joint at flexion angles of 30°–120° (load=500 N). High tensional stresses are observed at low flexion angles. The tensile stress in the notch considerably exceeds that in the humerus. With permission from Eckstein F, Merz B, Schön M, Jacobs CR, Putz R (1999) Tension and bending, but not compression alone determine the functional adaptation of subchondral bone in incongruous joints. Anat Embryol 199:85–97 (Copyright Springer 1999)

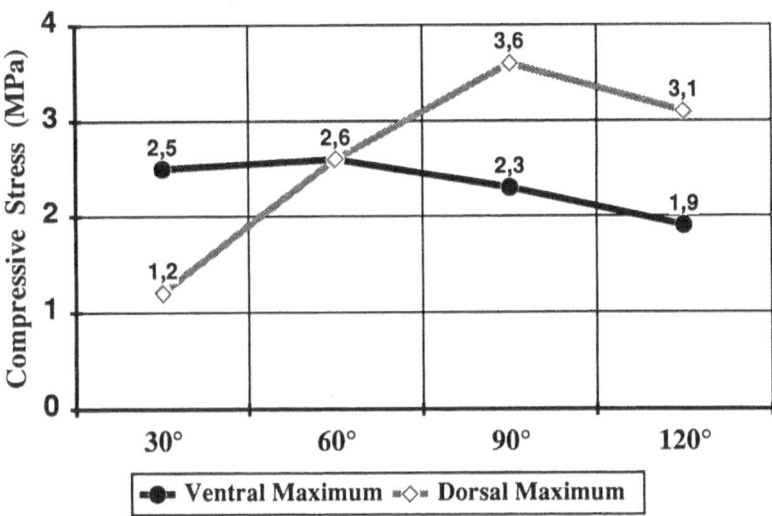

Fig. 36. Ratio of the ventral and dorsal pressure maximum of the trochlear notch in the anatomically based model of the humeroulnar joint; flexion angles of 30°–120°; load=500 N

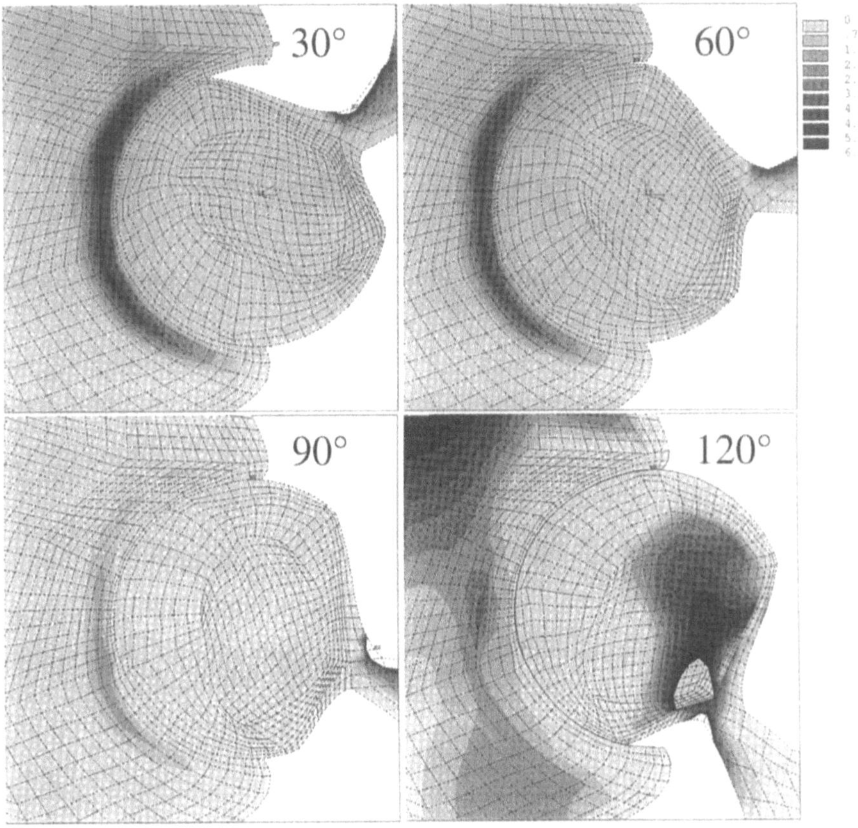

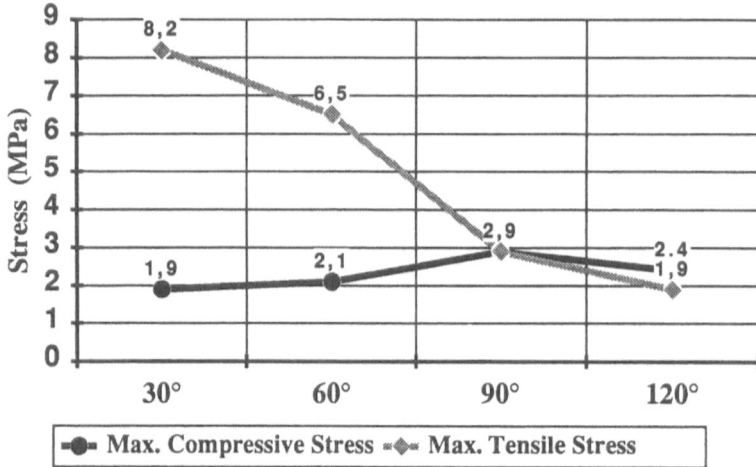

Fig. 38. Ratio of the maximal compressive vs. tensile stress in the subchondral bone of the trochlear notch in the anatomically based model of the humeroulnar joint; flexion angles of 30°–120°; load=500 N. With permission from Eckstein F, Merz B, Schön M, Jacobs CR, Putz R (1999) Tension and bending, but not compression alone determine the functional adaptation of subchondral bone in incongruous joints. Anat Embryol 199:85–97 (Copyright Springer 1999)

Table 1. Linear correlation (r) of mechanical output variables in the trochlear notch at a 30° and 90° flexion angle (load=500 N); distribution from ventral (coronoid process) to dorsal (olecranon)

	Subchondral compression	Subchondral tension	Subchondral SED
Flexion angle 90°			
Contact pressure	+0.98	–0.28	+0.58
Subchondral compression	–	–0.33	+0.56
Subchondral tension	–	–	+0.13
Flexion angle 30°			
Contact pressure	+0.97	+0.32	–0.07
Subchondral compression	–	+0.29	–0.11
Subchondral tension	–	–	+0.60

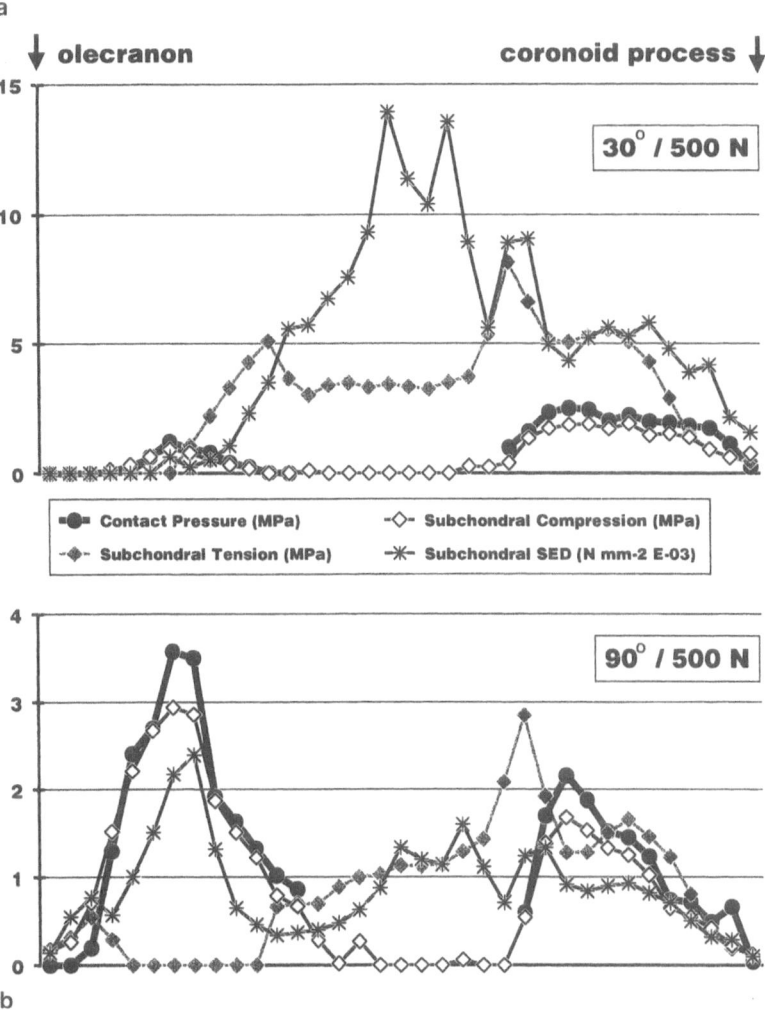

Fig. 39. Correlation of the contact stress, the compressive and tensile stress in the subchondral bone and the subchondral SED at 30° (**a**) and 90° (**b**) of flexion in the anatomically based model of the humeroulnar joint (load=500 N). With permission from Eckstein F, Merz B, Schön M, Jacobs CR, Putz R (1999) Tension and bending, but not compression alone determine the functional adaptation of subchondral bone in incongruous joints. Anat Embryol 199:85–97 (Copyright Springer 1999)

Figure 40 demonstrates the principal stresses acting throughout the trochlear notch. Only in the periphery of the joint, the stresses acting normal to its surface assume a considerable magnitude, whereas in the center the stresses acting tangential to the surface play a much more important role.

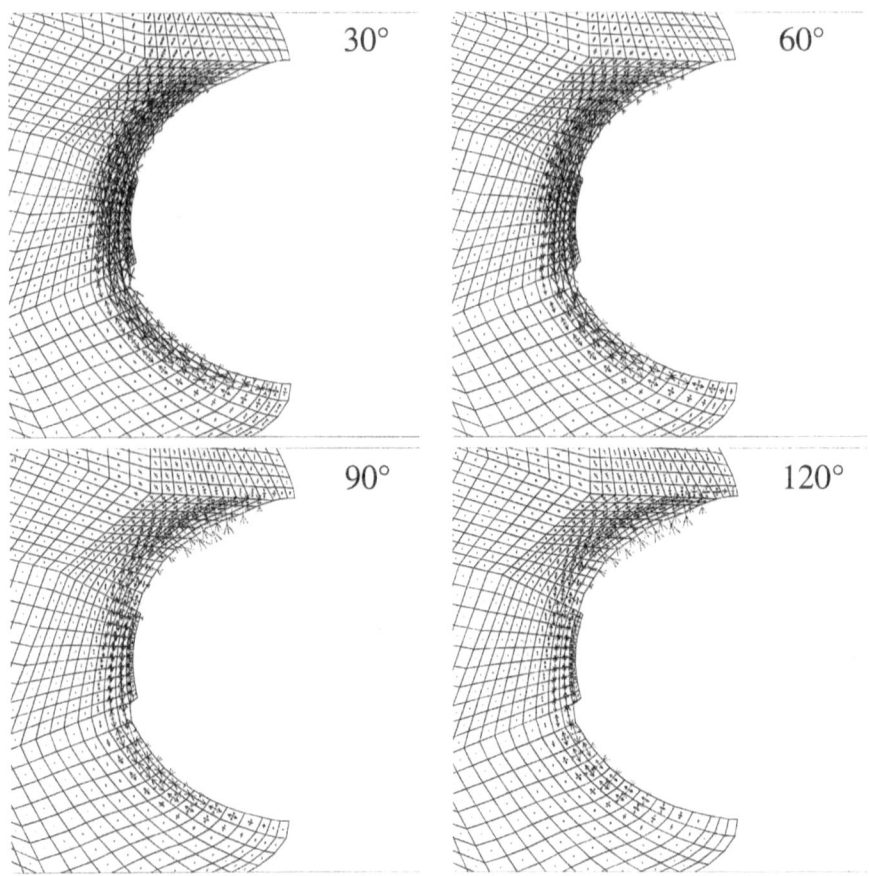

Fig. 40. First principal stress (mainly compressive, *arrows pointing toward each other*) and second principal stress (mainly tensile, *arrows pointing away from each other*) in the anatomically based model of the humeroulnar joint at flexion angles of 30°–120° (load=500 N). With permission from Eckstein F, Merz B, Schön M, Jacobs CR, Putz R (1999) Tension and bending, but not compression alone determine the functional adaptation of subchondral bone in incongruous joints. Anat Embryol 199:85–97 (Copyright Springer 1999)

4.4
Experimental Validation

The experimental determination of the contact areas in the same specimen revealed contact in the ventral and dorsal aspects of the articular surface at 250 N, but not in the depth of the notch. At 500 N a minimal joint space was still present, but at 750 N it had disappeared (Fig. 41). The contact areas thus became confluent between 400 and 600 N, both in the experiment and in the model.

Ventral and dorsal maxima of more than 2 MPa contact pressure were demonstrated by Fuji film at 250, 500, and 750 N, and with increasing force these slightly extended into the depth of the notch (Fig. 42). Here, no significant pressure

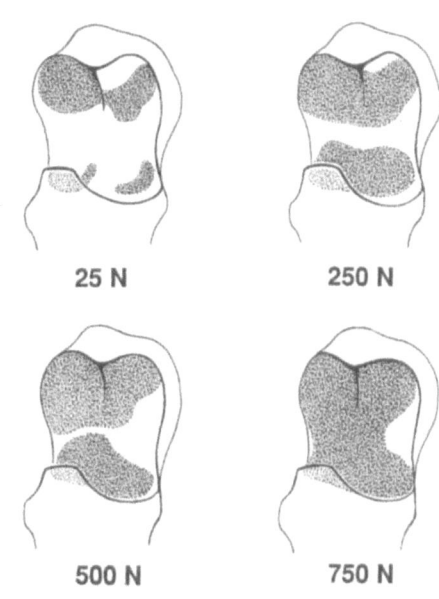

Fig. 41. Extension of the contact areas in the humeroulnar joint at loads of 25–750 N at 90° of flexion (experimental determination with polyether casts in the same specimen from which the FE model was constructed). With permission from Merz B, Eckstein F, Hillebrand S, Putz R (1997) Mechanical implications of humero-ulnar incongruity – finite element analysis and experiment. J Biomech 30:713–721 (Copyright Elsevier Science 1997)

25 N **250 N**

500 N **750 N**

Fig. 42. Pressure distribution in the humeroulnar joint at loads of 250, 500, and 750 N at 90° of flexion (experimental determination with Fuji-Prescale film in the same specimen from which the FE model was constructed). With permission from Merz B, Eckstein F, Hillebrand S, Putz R (1997) Mechanical implications of humero-ulnar incongruity – finite element analysis and experiment. J Biomech 30:713–721 (Copyright Elsevier Science 1997)

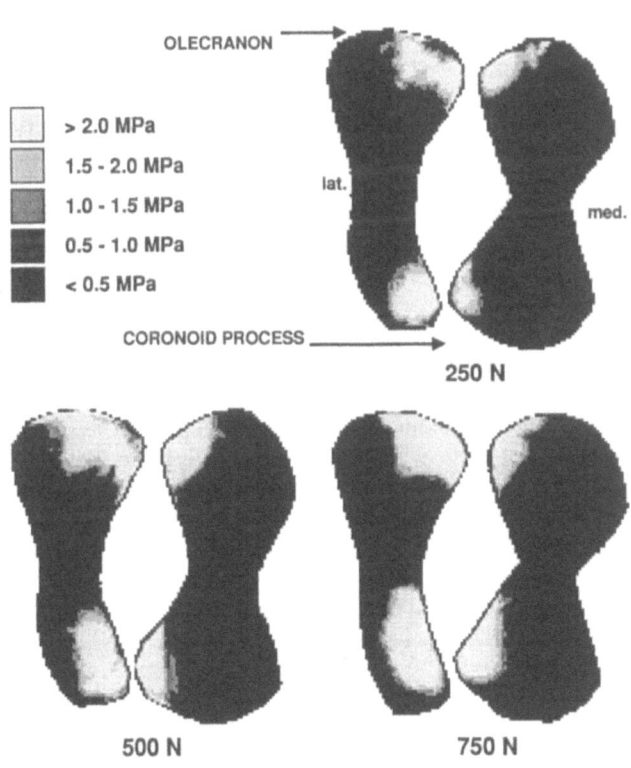

OLECRANON

> 2.0 MPa
1.5 - 2.0 MPa
1.0 - 1.5 MPa
0.5 - 1.0 MPa
< 0.5 MPa

lat. med.

CORONOID PROCESS

250 N

500 N **750 N**

(>0.5 MPa) was measured up to 500 N, as in the model. The size of the load-bearing areas (>2.0 MPa) in the experiment was similar to that in the simulation, although the load-bearing areas determined with Fuji film were located somewhat nearer to the edges of the articular surface than those in the model.

4.5
Morphological Findings at the Elbow Joint

4.5.1
Distribution of the Subchondral Mineralization

When being reconstructed from different data sets, the subchondral density distribution obtained by CT showed a high degree of reproducibility. From reconstruction to reconstruction 80% of 250 tested image points could be attributed to the same density interval by image analysis. Eleven percent of the image points deviated by more than 100 HU, 6% by more than 200 HU, and 3% by more than 300 HU.

Graphic summation of all 36 specimens produced a bicentric distribution of the subchondral bone density with ventral and dorsal maxima in the trochlear notch of between 800 and 900 HU. In the depth of the notch the subchondral value was about 100–200 HU lower (Fig. 43). In the fovea capitis radii, on the other hand, a central maximum was observed from which the values fell off concentrically towards the periphery. The density distribution of the humerus was similar to that of the opposing joint surfaces, although the difference between the patterns in the humeroradial and humeroulnar components of the joint was less marked than in the distal component. On average, the mineralization of the ulna and radius exceeded that of the humerus by about 100–200 HU (Fig. 43). In eight specimens the proximal and distal maxima lay within the same density interval, in 19 cases the mineralization of the distal component was 100 HU greater, in 7 cases 200 HU, and in 2 cases 300 HU greater than the proximal.

Obvious differences in the density distribution at the proximal ulna were found between the individual subgroups (type A, completely divided joint surface; type B, medial division of the joint surface; type C, continuous cartilage covering of the joint surface, Fig. 2). The bicentric distribution pattern was most marked in the completely divided joint surface. Here the ventral and dorsal density maxima exceeded the value in the depth of the notch by about 300 HU (Fig. 44a). A bicentric pattern was to be seen in the medially divided surface, but the maxima lay only about 100 HU higher than in the depth of the notch (Fig. 44b). In the continuous ulnar surface, a single central density maximum was observed from which the values first presented a plateau in the ventral and dorsal directions and finally tailed off towards the periphery (Fig. 44c).

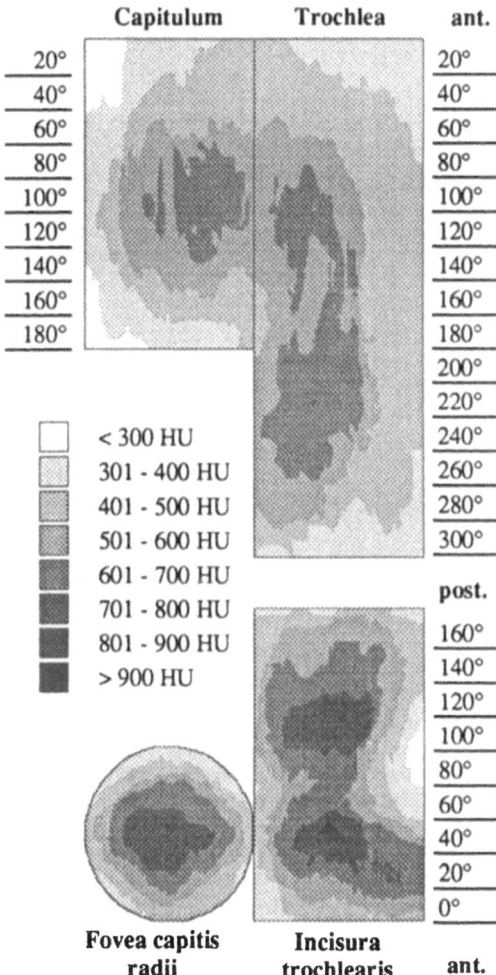

Fig. 43. Subchondral mineralization in the human elbow joint obtained with CT-OAM; average distribution pattern in 36 specimens (for the definition of the coordinate system of the joint template see Fig. 14). Whereas a bicentric (ventrodorsal) distribution of subchondral density is observed in the trochlear notch, a central maximum is recorded in the articular surface of the radial head. With permission from Eckstein F, Müller-Gerbl M, Steinlechner M, Kierse R, Putz R (1995) Subchondral bone density in the human elbow assessed by computed tomography osteoabsorptiometry: a reflection of the loading history of the joint surfaces. J Orthop Res 13:268–278 (Copyright Journal of Bone Surgery 1995)

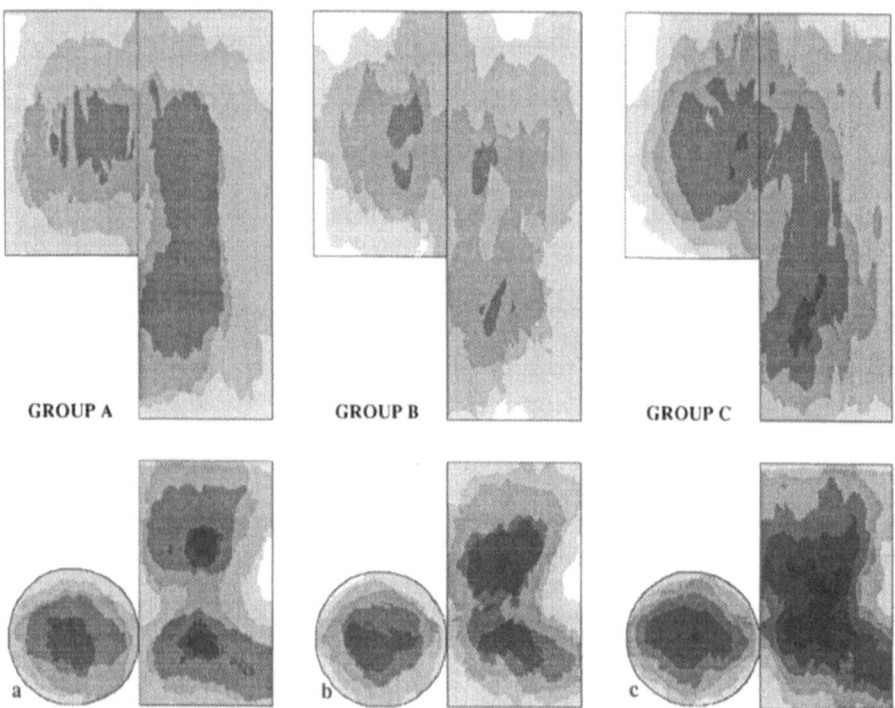

GROUP A   GROUP B   GROUP C

Fig. 44a–c. Subchondral mineralization in elbow joints with various types of surface morphology of the trochlear notch (obtained with CT–OAM); average distribution pattern of each subgroup (for an illustration of the respective type of surface morphology see Fig. 2). **a** Completely divided articular surface (16 specimens). **b** Medially divided articular surface (12 specimens). **c** Continuous articular surface (8 specimens). The bicentric distribution of subchondral bone density is more obvious with the divided articular surface. With permission from Eckstein F, Müller-Gerbl M, Steinlechner M, Kierse R, Putz R (1995) Subchondral bone density in the human elbow assessed by computed tomography osteoabsorptiometry: a reflection of the loading history of the joint surfaces. J Orthop Res 13:268–278 (Copyright Journal of Bone Surgery 1995)

4.5.2
Trabecular Bone Architecture

The trabeculae of the proximal ulna revealed a regular and typical orientation for the divided joint surfaces examined. Figure 45 shows a contact radiograph derived from a section of the specimen from which the model was made. A bony ridge can be seen in the depth of the notch which is characteristic of this type of articular surface. In the region of greatest subchondral density (ventral and dorsal to this) strong trabeculae run tangential to the surface; the dorsal ones continue towards the ulna tuberosity where the brachialis inserts. It is only near the periphery of the joint surface that trabeculae appear which are orientated perpendicular to it. These are finer and less dense, however, than the trabeculae in the zone of greatest subchondral density, which run tangential to the joint surface. This pattern was in principle observed in all the joints examined.

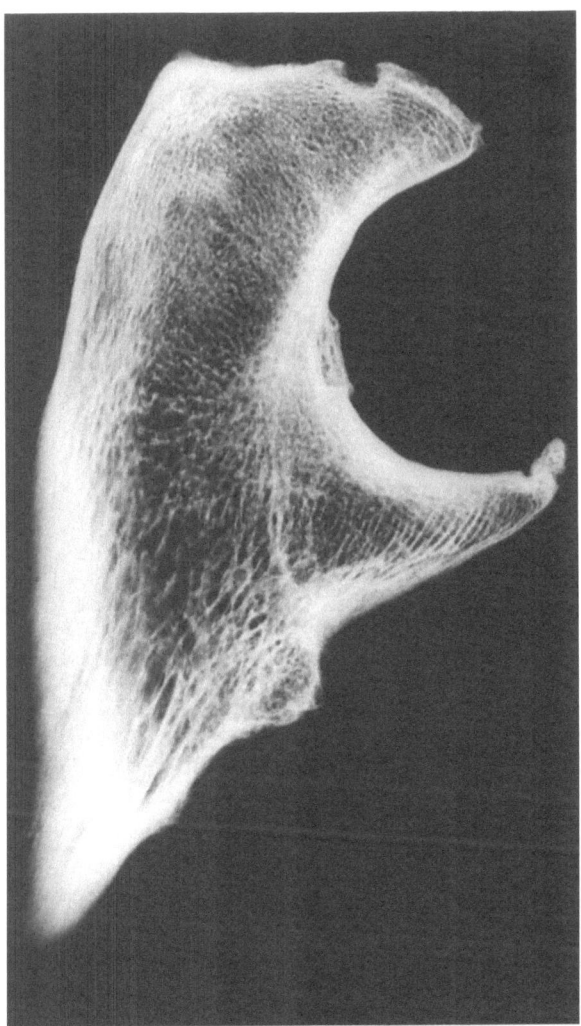

Fig. 45. Trabecular architecture of the trochlear notch (contact radiograph of a midsagittal section from the same specimen from which the FE model was constructed). Whereas in the periphery fine trabeculae are orientated normal to the articular surface, the stout trabeculae in the more central aspects of the joint surface and at the locations of the maximal subchondral density are aligned tangential to the joint surface. Please note the orientation of the first and second principal stress in Fig. 41. With permission from Eckstein F, Merz B, Schön M, Jacobs CR, Putz R (1999) Tension and bending, but not compression alone determine the functional adaptation of subchondral bone in incongruous joints. Anat Embryol 199:85–97 (Copyright Springer 1999)

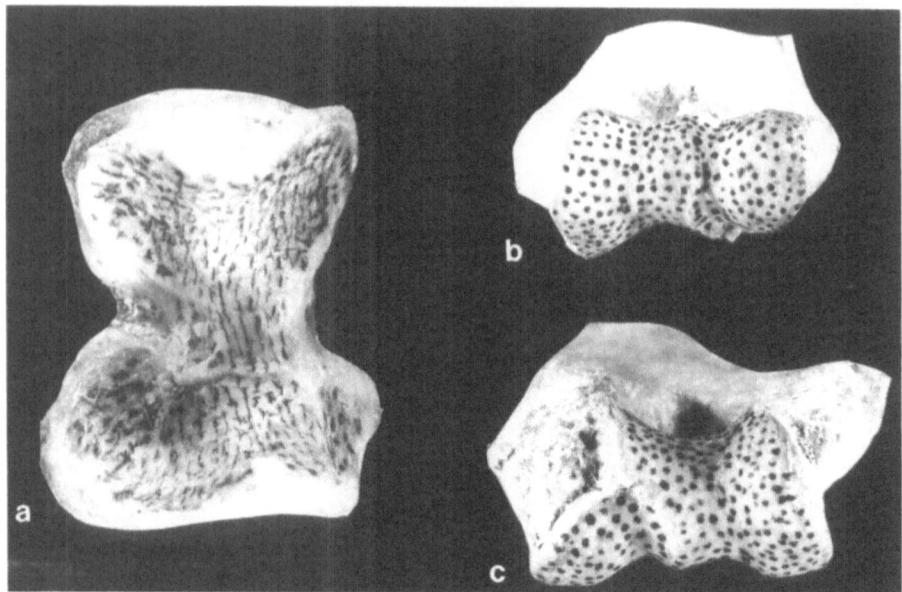

Fig. 46a–c. Preferential direction of the subchondral split lines in the humeroulnar joint. Whereas the split lines exhibit a clearly sagittal orientation in the trochlear notch, no such preferential direction can be observed in the trochlea humeri. With permission from Eckstein F, Merz B, Schön M, Jacobs CR, Putz R (1999) Tension and bending, but not compression alone determine the functional adaptation of subchondral bone in incongruous joints. Anat Embryol 199:85–97 (Copyright Springer 1999)

4.5.3
Subchondral Split Lines

The split lines of the subchondral plate of the ulna run in a predominantly sagittal direction (Fig. 46). Particularly in the region of the longitudinal ridge they were observed running from the olecranon, over the dorsal surface of the joint, the transverse bony ridge, and the ventral joint surface to the coronoid process. Only in the medial and lateral periphery of the articular surface did they occasionally show a less clear orientation. The split lines of the humeral joint surface (Fig. 46), on the other hand, showed in most cases no preferential orientation. Only in a few cases did they run transversely from the medial edge of the trochlea, over the capitulum to the lateral epicondyle.

5 Discussion

5.1
Methodological Discussion

The method of computer simulation is particularly suitable for the analysis of the pressure transmission in joints and for the functionally adaptive processes of the connective tissues because the mechanical stresses and strains can be calculated for locations that are not accessible to direct artifact-free measurement (for instance, the subchondral bone). Single factors, which cannot be modulated experimentally (such as fine differences in congruity/incongruity), can be isolated and varied in a complex system and investigated parametrically. Furthermore, biological processes such as the remodeling of bone, which would take up months of an experimental investigation, can be tested efficiently within reasonable periods of time.

As noted above, however, the validity of the information obtained from a computer simulation depends strongly upon the assumptions on which the model is built. This includes the quantitative distribution of the tissue (e.g., cartilage thickness), its material properties (e.g., the elastic modulus) and the mathematical formulation of the adaptive processes. Since, for technical and practical reasons, the modeling involves certain simplifications which are unavoidable, it is necessary to confirm the predictions of the model and to check their plausibility within the context of the experimental and morphological findings available. The following methodological discussion critically evaluates the basis of the computer models and the methods employed for obtaining the experimental and morphological data.

5.1.1
Design of the Computer Models

5.1.1.1
Joint Incongruity

In the idealized model, the type and magnitude of the incongruity were taken from previous experimental work, and in the anatomically based model, from a biomechanical experiment performed in the course of this study. A polyether cast was used to reproduce the width of the joint space under the minimal load necessary to bring the joint components into contact. The thickness of the cast was then measured with a spherical sensor perpendicular to the surface of the convex joint component.

The cast material employed (Permadyne, ESPE, Seefeld) is used in dental practice and has an excellent capacity for reproducing shape. The casts were kept at normal room temperature and measured within a week; the preservation of their geometric accuracy for this time was guaranteed by the manufacturer. The casts were obtained under a load of 25 N; this force was kept constant for the 1–2 min that the material takes to set. This force is necessary to squeeze the material out of the space between the joint components and to bring them together. Since it produces only relatively small contact areas, it can be assumed that there is only a minimal degree of deformation of the articular cartilage. The thickness of the cast was measured with a sensor weighing 30 g. Because of the consistency of the cast material this is unlikely to produce any significant deformation of the cast by the measuring apparatus, the reproducibility of the methods having been confirmed in a previous study (Eckstein et al. 1993).

The force was applied by simulating the action of the triceps. This allows the examination of various angles of flexion and the natural contact situation of the joint components without constraints. Simulation of the flexor muscles was neglected, since here several muscles contribute to compensate the rotational moments. Such an experimental set-up would not only be technically difficult, but it would also constitute a mathematically indeterminate problem (An et al. 1984; Brand et al. 1994) since it cannot be determined with certainty what force should be taken for the individual flexor muscles (brachialis, biceps, brachioradialis). From the design of our experiment it emerged that by imitating the action of the triceps at a flexion angle of 90° of the elbow joint, the joint reactive force passed roughly through the center of the trochlear notch. This corresponds to the direction that other authors have calculated for the conditions under flexor activity (Pauwels 1963, 1965, 1980; An et al. 1984). It is theoretically conceivable that the application of the load at another site (or muscle insertion) could produce a different kind of deformation of the bone, but this is not to be expected with a force of 25 N. It can therefore be assumed that the simulation of flexion against resistance would produce nearly identical results so far as the physiological incongruity of the humeroulnar joint is concerned. For these reasons we can assume that the measured thickness of the cast reliably reflects the width of the joint space in the unloaded situation of the joint components. In this connection it is also of interest that the incongruity could be qualitatively confirmed by high-resolution MRI in joints with intact capsule and ligaments (Eckstein et al. 1996a).

It should be noted that these values refer to the relative incongruity of the joint components and not to their absolute shape. The analysis of absolute shape requires alternative methods, such as stereophotogrammetry (Ateshian et al. 1991, 1994a). With this type of technique, however, combining the data sets from the opposing joint components and determining the relative deviations in shape is accompanied by several methodological problems and artifactual pitfalls. Since the relative variations in shape predominantly affect the distribution of pressure throughout the joint surface, the direct measurement of the width of the anatomical joint space has been given preference in the present investigation. For the construction of the model it was assumed that the joint surface of the humeral trochlea is circular in sagittal sections, which has been substantiated by previous morphological examinations (Shiba et al. 1988) and that, with regard to the incongruity at various angles of flexion, no significant differences have been reported in the humeroulnar joint (Eckstein et al. 1995a).

It should be emphasized that the two-dimensional models described here are limited to the analysis of the effects of joint incongruity in a single plane. Since the humeroulnar articulation is similar to a hinge, the movement of which can be essentially represented in one plane, this simplification appears acceptable. Moreover, the concave incongruity can in principle be described in every sagittal section through the humeroulnar joint, the typical bicentric density distribution being found in all sagittal sections through the trochlear notch, and, of particular importance, the predictions from the two-dimensional model being confirmed in a three-dimensional biomechanical experiment. The basic question of this investigation could therefore be answered on the basis of the two-dimensional models and these permit analogous conclusions about the effects of concave and convex incongruity on other human joints. It should be noted, however, that in joints such as the hip two-dimensional models cannot adequately characterize the pressure distribution and functional adaptive processes, and in these cases three-dimensional modeling is essential. The formulation of three-dimensional nonlinear contact algorithms in FE analysis, however, is a complex problem, and is presently still in its developmental stage.

5.1.1.2
Cartilage: Quantitative Distribution, Material Properties, and Nonlinear Contact Behavior

The idealized joint model assumed a simplified uniform thickness of the cartilage. This was based on mean values reported for the elbow joint (Kurrat and Oberländer 1981; Oberländer and Kurrat 1982, Schenck et al. 1994; Milz et al. 1997; Springer et al. 1998). Since the principal goal of this part of the investigation was to selectively and parametrically analyze the effect of various types and degrees of joint incongruity on the pressure transmission and functional adaptation of the bone tissue, a homogeneous distribution of the thickness in these models was chosen. However, in reality the cartilage tissue in the elbow joint is not uniformly distributed, and in the majority of cases no cartilage is found in the depth of the adult trochlear notch (Tillmann 1971, 1978; Oberländer and Kurrat 1982; Schenck et al. 1994; Milz et al. 1997; Springer et al. 1998). This was taken into account in the anatomically based model, since the inhomogeneous thickness of the cartilage of a specific joint was determined by image analysis and used in designing the model. Contemporary developments in the field of MRI, which now make it possible to measure the thickness of the cartilage noninvasively in the living, are described in Sect. 5.5.1.

While the cartilage thickness in the present investigation was regarded as a given and not as a dependent variable, the question arises whether functional adaptation of the cartilage should be taken into account in the modeling process. This is of interest because a change in its thickness may also be reflected in the shape of the articular surface, and will therefore alter the "fit" of the joint. In accordance with the theories of Pauwels (1960, 1965, 1980) and Kummer (1963), the thickness of the cartilage has often been regarded in the anatomical literature as the reflection of the long-term mechanical stress acting on the joint surface (Tillmann 1971, 1978; Kurrat and Oberländer 1978; Müller-Gerbl et al. 1987; Eckstein et al. 1992; Milz et al. 1995, 1997; Adam et al. 1998). Computer simulations of the pressure transmission in synovial joints, a direct relationship between the local hydrostatic pressure and the cartilage thickness has also been postulated, while shear stress has been assumed to promote the progress of calcification to the joint surface (Carter 1987; Carter et al. 1987a,b, 1991; Carter and

Wong 1988; Wong and Carter 1990; Smith et al. 1992). The results of animal experiments have suggested that the cartilage thickness diminishes during immobilization (Jurvelin et al. 1986; Helminen et al. 1992). They have also indicated that, following moderately long-term mechanical loading, the cartilage thickness increases slightly (Kiviranta et al. 1987, 1994; Helminen et al. 1992). Cell biological investigations have also demonstrated that, with moderate dynamic loading, the synthetic activity of the chondrocytes increases (Carters and Lowther 1978; Sah et al. 1989; Urban 1994; Bachrach et al. 1995). It is not yet known, however, to what extent a functional adaptation of the cartilage thickness to mechanical loading takes place in adults. A recent comparison of cartilage thickness values in healthy triathletes and physically inactive individuals has shown that there exists a great degree of interindividual variability in knee joint cartilage thickness, but that there are no significant differences between these two (Mühlbauer et al. 1998, 1999). In particular, one is still far from being able to formulate these relationships in a way suitable for mathematical modeling. For these reasons in the present investigation we have avoided a simulation of the differentiation and functionally adaptive processes of the articular cartilage. Such investigations should, however, help in the future to reveal the interesting interaction between mechanical stresses and strains, the differentiation of the connective tissue and the fine modeling of the shape of the joint surfaces (see Sect. 5.5.5).

Regarding the representation of the material properties of the cartilage in the model, the suggestions in the literature are not consistent. It is clear that articular cartilage is a multiphasic tissue, and that its deformational behavior is time dependent. Under compression three phases can be differentiated in the response of the cartilage: initial deformation, a viscoelastic phase, and equilibrium. During the initial phase the response of articular cartilage is virtually linear and elastic, and it is nearly incompressible (Hayes et al. 1972; Mow et al. 1989; Clift et al. 1992; Goldsmith et al. 1996). This does not mean that no deformation takes place, but that the total volume of the cartilage remains the same under local compression. Such a response can be represented by a Poisson number of almost 0.5 (in our model we have selected 0.495–0.499). The elastic modulus is required as a further quantity; however, its analysis is problematic. For one thing, the initial stiffness of the tissue is affected by the speed of the loading (dynamic stiffening; Oloyede et al. 1992; Kim et al. 1995), and, for another, the measured stiffness depends upon the exact moment at which the deformation is determined. This is because immediately after load application the fluid in the articular cartilage starts to flow, and viscoelastic deformation begins. A further complication follows from the fact that investigations into the deformation of cartilage have sometimes been undertaken in confined compression, when the cartilage cannot expand laterally and the collagen fibrils cannot contribute to the stiffness of the tissue as they would under natural in situ conditions. Compression tests with artificial indentors have shown that the initial elastic deformation of the articular cartilage which is strongly affected by the collagen content, and which, contrary to long-term static loading, the proteoglycans are of only subsidiary importance here (Mizrahi et al. 1986; Jurvelin et al. 1988). Determining the material properties from indentation tests, however, is complex and, again, requires mathematical modeling with inherent simplifications. For these reasons the elastic modulus used in this investigation was taken from the work of Kim et al. (1995), where the dynamic stiffness of cartilage was determined in unconfined tissue samples at frequencies of 0.001–1 Hz. With frequencies between 0.1 and 1 Hz, which seem to represent the physiological situation realistically, a relatively constant elastic modulus of about 15 MPa has been described. This value is also supported by recent investigations in which it has been possible to measure the initial deformation of cartilage under loading for a period of only 150 ms (Shephard and Seedholm 1996). This yielded values in the region of 5–40 MPa.

To check the sensitivity of the bone remodeling simulation in terms of the cartilage material properties in the idealized model, the value was systematically varied between 5 and 25 MPa. No important effect of the cartilage stiffness on the distribution pattern of the bone density was established. Systematic variation was not applied to the anatomically based model, because, in comparison with the biomechanical experiment, very similar behavior of the contact areas and an almost identical contact pressure was established with an elastic modulus of 15 MPa. Since deformation of more than 20% was not observed in any of the models, the choice of an infinitesimal material model appeared justified. With greater loads finite deformation should be taken into account mathematically. It must be remembered that the material properties of the articular cartilage can also vary throughout the joint surface (Athanasiou et al. 1991; Schenck et al. 1994). Owing to the poverty of available data, such local differences were not modeled in the present investigation. As the parametric analysis of the idealized models has shown, however, local differences in cartilage material properties should have only negligible effects on the adaptation of the subchondral bone density.

If the loading of the articular cartilage is maintained for some length of time, its deformation depends strongly on the fluid flow (viscoelastic behavior). This mechanical behavior can be approximated by the biphasic theory of mixtures (Mow et al. 1980, 1984; Setton et al. 1993; Mow and Ratcliffe 1997); it is highly time-dependent, and there is no linear relationship between stress and strain. For a certain time, the load is supported mainly by the fluid phase of the articular cartilage, in which an interstitial hydrostatic pressure is building up (Soltz and Ateshian 1998). During this period the "solid" proteoglycan-collagen matrix is substantially protected from mechanical deformation and elastic stress (Ateshian et al. 1994b; Ateshian and Wang 1995; Kelkar and Ateshian 1995). If the fluid continues to flow, the hydrostatic pressure is continuously reduced until equilibrium is reached, and the entire load is carried by the solid matrix. This equilibrium value has been determined in "creep" experiments (prolonged application of a constant load) in tissue samples under confined compression (aggregate modulus) and differs considerably from the elastic modulus (which is determined in unconfined compression). The aggregate modulus shows a linear relationship with the applied load and lies characteristically between 0.5 and 1 MPa. This value, however, is a measure of the long-term deformation of the tissue under confined compression in the creep experiment and must not be equated with the initial deformation during physiological loading.

The material constants on which the present investigation has been based, however, are related to the initial deformation of the tissue during cyclic loading, such as predominates during daily use of the joint. It does not apply to the mechanical behavior of the articular cartilage under long-term static loading, which would require a biphasic formulation of the cartilage tissue properties. A biphasic mathematical formulation for the cartilage has been developed and implemented with the FE method (for reviews see Clift et al. 1992; Goldsmith et al. 1996), but the corresponding algorithms for the nonlinear contact of the two biphasic cartilage layers is still in the developmental stage (Donzelli and Spilker 1993, 1995, 1998). With these formulations it would also be possible to consider the effect of the joint fluid. Preliminary biphasic analyses of pressure transmission in concavely incongruous joints, undertaken in collaboration with Peter Donzelli and Bob Spilker (Rensellear Polytechnic Institute, Troy, New York), are reported in Sect. 5.5.2. Further developments in the FE method can make it possible to also carry out time-dependent analyses of long-term static loading of incongruous joints, taking into account the synovial fluid. Most importantly, they would allow – unlike a linear elastic model – information to be obtained about the load partitioning between the solid phase (proteoglycan-collagen matrix) and the interstitial fluid phase (water and ions). They would therefore permit a more detailed analysis of the etiology of mechanically induced degeneration of articular cartilage.

5.1.1.3
Bone Tissue: Quantitative Distribution, Material Properties, and Functional Adaptation

Since this investigation focused principally on the subchondral bone, the external shape of the bone during the remodeling simulation in the idealized model was predetermined and not regarded as a variable quantity. That means that the "internal" remodeling, but not the "external" modeling (or surface remodeling) was simulated. Since in the idealized model we sought to determine the specific effect of the joint shape on the density distribution within the subchondral bone, changes in the external form of the bone were not desired. A simulation of external modeling is technically demanding, because each of the nodes of the FE grid must be dynamically adapted and relocated during the simulation. In their simulation of bone remodeling in the proximal femur, Beaupre et al. (1990b) also dispensed with the external remodeling because in their opinion the resulting changes would be minimal. However, depending on the specific questions asked – prediction of the long-term action of endoprostheses, for instance – it is under certain circumstances necessary to take changes in the shape of the bone into account (see Weinans et al. 1993; van Rietbergen et al. 1993). In connection with the present investigations, the simulation of the external remodeling processes is of importance when the differentiation and growth during the ontogenetic development of joints are to be included in the simulation. This, however, would require a much more advanced mathematical theory of the processes of connective tissue differentiation, growth, and functional adaptation than is presently available. Contemporary aspects of this subject are discussed in Sect. 5.5.5.

The results of the bone remodeling simulation in the idealized model suggest that not only compression but also tensional stresses and strains make an important contribution to the functional adaptation of the subchondral bone tissue in incongruous joints. These tensional stresses arise as a result of the bending of the bone during load transmission. To what extent a bone is subjected to bending, however, depends largely on its internal structure and external shape. Since, in the idealized models, for instance, no cortical bone is represented which could provide the joint components with stability and limit bending, this hypothesis was tested on the anatomical model. In this specific model of the humeroulnar joint, no homogeneous distribution of the bone tissue was assumed, but the local stiffness of each was calculated from the local Hounsfield values of the CT data sets. It is worth remarking that stiffness and density must not be unconditionally treated as synonymous, but there is a high correlation between these two entities (Carter and Hayes 1977; Gibson 1985; Rice et al. 1988; Hodgkinson and Currey 1990; Lotz et al. 1990; Rho et al. 1995). However, the additional effects of microstructural parameters which may for instance be summarized under the term "fabric tensors" (van Rietbergen et al. 1996a) also play an important role here. Various morphological and micro-CT methods now make it possible to derive such microstructural parameters (Feldkamp et al. 1989; Kuhn et al. 1990; Müller et al. 1994, 1996, 1998; Müller and Rüegsegger 1995; van Rietbergen et al. 1995; Rüegsegger et al. 1996; Odgaard 1997; Graichen et al. 1999). Since, however, the percentage influence of the density on the stiffness has been estimated to range from 72% (Hodgkinson and Currey 1990) to 90% (Lotz et al. 1990), a CT-based model allows a relatively accurate evaluation of the local bone stiffness.

Since the tubular architecture of long bones and the connection of the subchondral bone with the cortical shell cannot be adequately represented in a two-dimensional model, a so-called "side plate" was introduced into the anatomically based model. The side plate is mechanically coupled to the actual model (the "front plate") and adds additional stabilization, thus increasing the reliability of the predictions in the two-dimensional case (Verdonshot and Huiskes 1990). The "side plate" was conceived in terms of the anatomical relationships of the humeroulnar joint. The good agreement between the contact behavior and joint pressure in the model and in the biomechanical experiment suggests that the three-dimensional structure of the joint was realistically represented in the model. With regard to the tensional stresses in the subchondral bone it is of crucial importance that the prediction of the model could also be substantiated by examination of the trabecular architecture and the split lines of the subchondral plate.

In the idealized model, a uniform distribution of the bone density and stiffness was assumed at the beginning of the simulation. During the course of the simulation this was adapted depending on the local stresses and strain based on the bone remodeling theory employed (Beaupre et al. 1990a,b; Jacobs et al. 1995). The comparative analyses showed that the predictions were only to a limited extent affected by the distribution of bone density at the beginning of the simulation. The predictions of the model also agree with the morphological findings on the subchondral mineralization of joints with a corresponding "fit."

Our simulations were based on the SED theory of Beaupre et al. (1990a,b), the fundamental theoretical and biological background as described in detail above (see Sect. 2.3). As outlined there, the cellular mechanism of the biomechanical coupling are still only partially understood, and the actual mechanical stimulus for bone remodeling has not been determined with complete certainty. One reason for the usefulness of the SED is its scalar (nonvector) nature as a feedback signal, and it is also used in other contemporary theories of bone remodeling (Weinans et al. 1989, 1993; van Rietbergen et al. 1993). We therefore assume that these alternative theories would have produced very similar (if not identical) results. As has been demonstrated by Carter et al. (1987c), the use of diverse mechanical failure criteria or microfractures as feedback signals lead to similar mathematical formulations, and for these no other results are therefore to be expected.

On the basis of recently published suggestions that the fluid flow in the canalicular network is registered by the osteocytes and translated by the osteoblasts and osteo-clasts into a biological response (see Sect. 2.3.2), one is inclined to consider whether the strain gradients rather than the absolute values of the strains themselves, or the SED, should not be used in a simulation. At the present time there is no mathematical theory that would allow the implementation of these relationships with the FE method. It is also questionable whether such a simulation would produce significantly different results, since the regions with the highest strain gradients are in close vicinity to those of the highest absolute strain and SED values. It must still be deter-mined whether the effects of varying strain rates (Turner et al. 1995) and strain frequencies (Turner et al. 1994; Rubin and McLeod 1995) are of any particular signifi-cance. Although further refinements of the existing bone remodeling theories are to be expected, particularly in view of new discoveries in cell biology, it must be noted that the presently available SED-based theories have been able to deliver predictions which are in agreement with experimental findings such as the present simulation and with clinical observations.

Although in the theory employed the calculation of the local stresses takes into account that bone has a porous structure, and that the stresses in mineralized tissue are higher than at the continuum level, changes in density are eventually computed at the latter. FE models have recently been based on small samples of trabecular bone in which single trabeculae have been represented (Beaupre and Hayes 1985; Hollister et al. 1991; Müller and Rüegsegger 1995; van Rietbergen et al. 1995). The computational demands, however, are considerable, particularly for structures of the order of whole bones. Bone remodeling simulations at the microstructural level are, however, cur-rently emerging (e.g., van Rietbergen et al. 1996; Mullender et al. 1998), and it remains to be seen to what extent the information gained will increase our understanding of the functional adaptation of subchondral bone.

5.1.1.4
Load Application and Boundary Conditions of the Models

In the idealized models the load application was selected in agreement with biomechanical analyses for strenuous, moderate and light activity (An et al. 1984; Morrey 1992; Donkers et al. 1993). In the anatomically based model, the applied loads exceeded what one would expect, for instance from slow push-ups (Donkers et al. 1993). They should thus cover a realistic range of physiologically relevant loads. The direction of the force was obtained both in the simulation and in the biomechanical experiment from the equilibrium between the tension exerted by the triceps and that of an external load. The boundary conditions of the anatomically based model there-fore corresponded to static loading of the humeroulnar joint under extension against resistance for each of the positions 30°, 60°, 90°, and 120°. It is conceivable that with greater loads the point of insertion of the muscle plays a role for the strains and particularly the bending occurring in the bone. With a dorsally directed joint reaction force, for instance, the olecranon is stabilized in the simulation by the pull of the triceps, whereas with a ventrally directed force the coronoid process receives no such bracing. It is therefore possible that with flexion against resistance (as occurs when carrying a heavy load), the strains in the subchondral bone are differently distributed.

The simulation of such loading was not examined in the present investigation, but it could in principle be analyzed in the model presented here, if the boundary conditions were to be altered and the local strains and stresses calculated anew.

5.1.2
Biomechanical Experiment

The load application and boundary conditions in the biomechanical experiment have been described above (see Sect. 5.1.1). To guarantee a high degree of comparability with the simulations these were defined almost identically to those in the model. The method for measuring the width of the joint cavity under minimal force by means of polyether casting material has also been discussed above. The determination of the contact areas under greater force was undertaken in the same way and with the same casting material as the measurement of the width of the joint space.

The position of the contact areas, however, does not permit conclusions to be drawn about the pressure distribution within the joint surface, since an inhomogeneous pressure within the contact areas cannot be demonstrated. The pressure at the joint surface can be measured directly using Fuji Prescale film (Fukubayshi and Kurasowa 1980; Hehne 1990). With correct use and calibration this can provide values that agree with those of other methods (Ateshian et al. 1994b). By employing one strip for the medial and one for the lateral articular surface it was possible to eliminate crinkle artifacts. It is, however, particularly important to remember that the thickness (about 0.1 mm) of the film itself can affect the "fit" of the joint. This is expressed in the more peripheral location of the load-bearing areas compared with the computer simulation and the contact areas determined by the casting material. Nevertheless, a combination of the two methods allows a reliable experimental analysis of the pressure transmission in the joint.

Both the merging of the ventral and dorsal contact areas under a force of about 500 N and the bicentric distribution of the pressure (with values over 2 MPa in the dorsal and ventral parts of the ulnar surface and values smaller than 0.5 MPa in the depths of the trochlear notch) were found both in the experiment and in the computer simulation. This agreement supports the validity of the anatomically based FE model of the humeroulnar joint.

5.1.3
Morphological Investigations on the Elbow Joint

The subchondral bone is not accessible to artifact-free measurement of the mechanical stresses and strains with currently available methods. Therefore the predictions of the model were compared with the results of morphological investigations.

5.1.3.1
Determination of the Subchondral Density
with CT Osteoabsorptiometry

CT-OAM makes it possible to determine the radiological density of the subchondral bone and to provide information about its mineralization and calcium content noninvasively. The technical and methodological basis of this procedure has been investigated and presented in detail in earlier publications (Müller-Gerbl et al. 1989, 1992; Müller-Gerbl 1998). Specifically with regard to the analysis of the elbow joint we found a high reproducibility by using section intervals of 2 mm from different data sets. Three-dimensional CT-OAM (Müller-Gerbl et al. 1992; Müller-Gerbl 1998), by which the form of the bone can be three-dimensionally reconstructed from CT data, has the advantage that individual joints can be visualized in their original form without distortion. On the other hand, the method of reconstruction on the basis of joint surface templates used here provides a mean distribution pattern for all specimens and subgroups which facilitates an inclusive and objective judgement.

As noted above, the subchondral density distribution pattern of the idealized joint models can be confirmed by the mineralization pattern found by CT-OAM. This supports the predictive value of the bone remodeling theory employed (Beaupre et al. 1990a,b; Jacobs et al. 1995) and also shows that the corresponding mineralization pattern can be obtained by starting from a homogeneous distribution based solely on a specific mechanical situation brought about by joint incongruity.

5.1.3.2
Assessment of the Trabecular Architecture

The trabecular bone presents a porous structure of which the microarchitecture is characterized by a complex arrangement of thin (about 100–200 µm) strong plates and struts or rods (Raux et al. 1975; Singh 1978; Hayes and Snyder 1981; Odgaard 1997). The bony structures recognizable in a single section through this network – the trabeculae – give a qualitative impression of the alignment of the plates and rods in trabecular bone.

As early as the end of the nineteenth century, Wolff (1892) suggested that the trabeculae are lined up by a process of adaptation along the principal stresses. As explained above, the simulation of bone remodeling used in this work allows its density to be predicted, but not its microarchitectural arrangement. The investigation has, however, established substantial agreement of the first principal stress (mainly compressive) and the second principal stress (mainly tensile) in the static analysis of the idealized model with the orientation of the trabeculae in the trochlear notch. This correlation supports the prediction of tensional stresses tangential to the articular surface in the proximal ulna.

New developments in the bone remodeling theory (Jacobs 1994; Jacobs et al. 1997, 1998b) also allow the prediction of the anisotropic material properties of the tissue in the remodeling simulation, i.e., direction-dependent stiffness. This type of research could refine the analysis of the interdependence between mechanical loading and trabecular microarchitecture which is to be dealt with in Sect. 5.5.3.

5.1.3.3
Analysis of the Subchondral Split Lines

The examination of split lines, which were first observed in the skin, was undertaken by Hultkrantz (1898) with the aim of analyzing the direction-dependent tensile strength of the surface layer of articular cartilage. He suspected that the split lines have the same orientation as the collagen fibrils and attributed their direction to the tensional stresses in the tissue caused by friction and by pressure. Recent ultrastructural (Jeffery et al. 1991) and biomechanical (Akizuki et al. 1986) investigations have confirmed the relationship between the course of the split lines, the preferential direction of the collagen fibrils, and the maximal tensile stiffness.

Benninghoff (1931) and others used the split line method to investigate the architecture of compact bone in decalcified specimens (for a recent summary see Vogt et al. 1999). Until recently, however, there have been no studies on the split lines in the subchondral bone plate.

The sagittal alignment of the subchondral split lines of the trochlear notch, which were also observed in the surface layer of the cartilage by Hultkrantz (1898) and Tillmann (1978), leads to the conclusion that the bone tissue is indeed adapted to the tensional stresses, as is predicted in this plane by the models. Split lines occasionally appear in the subchondral plate of the trochlea, and are transversely orientated in the same way as in the surface layer (Hultkrantz 1898; Tillmann 1978). This also agrees with the predictions of the models in the sense that they do not show tensional stresses in the sagittal plane.

5.2
Effect of the Incongruity on the Pressure in the Joint

5.2.1
General Observations

The stress distribution throughout the surface of congruous spherical joints has been previously analyzed by Kummer (1968), Kummer et al. (1987), and Mockenhaupt (1990). They emphasized that no uniform pressure is to be expected in such joints, but that with a central joint reaction force it decreases from a central maximum towards the periphery of the socket. This is because, with freedom from friction, only those partial vectors can be transferred that are oriented parallel to the applied force. Whereas the local normal vectors (perpendicular to the joint surface) contribute to the transmission of the load, the tangential vectors (parallel to the articular surface) rise up in pairs in both halves of the joint.

It has been suggested that a central stress peak is to be expected with a less curved socket (convex incongruity) because of the relatively small contact areas. This has been supported by the photoelastic studies (Kempson et al. 1971; Tillmann 1978) and by computer simulations (Mockenhaupt 1990; Eckstein et al. 1994a, 1995b, 1996a). Tillmann (1978) has shown that by removing the central region in a congruous photoelastic model, a bicentric distribution of the contact stress can also be achieved.

This geometric configuration is not, however, directly comparable to that of a deeper socket.

So far as the pressure transmission in concavely incongruous joints is concerned, Kummer (1974, 1985), Goodfellow and Mitsou (1977), Bullough (1981), and Greenwald (1991) suggested that a "load-distributing" effect is to be expected, and that the maximum surface pressure is here less than in an ideally congruous joint. It was assumed that such a primarily incongruous joint only becomes secondarily congruous under loading, but that the periphery carries more load than in the congruous case. We were able to confirm this in previous FE models (Eckstein et al. 1994, 1996a), but these findings should be taken with caution, since no articular cartilage was modeled. Simulations based on linear elastic material models of the cartilage (Eckstein et al. 1995b) have shown that the contact pressure in the concavely incongruous joint lies, with greater loading and increasing contact in the joint, considerably below the values found in a primarily congruous joint. The models of concave incongruity used in this investigation also show the peripheral location of the load-bearing surfaces and the associated load distributing effect. It must nevertheless be pointed out that conclusions about the mechanical stress acting on the articular cartilage, and therefore about the possible etiological factors of mechanically induced osteoarthrosis, should be drawn with some caution. The cartilage is a multiphasic tissue, the loading of which is shared between the solid collagen-proteoglycan matrix and the fluid phase. Loading first leads to an increase in the interstitial hydrostatic pressure because the solid matrix presents considerable resistance to fluid flow from the contact areas (Mow et al. 1980, 1984; Ateshian et al. 1994b). This causes the solid matrix to be at first subjected to very little strain and elastic stress, as the tissue is able for a short time to withstand very high pressures without suffering from morphological damage. The distribution of the load between the matrix and the interstitial fluid constitutes a very important "protective" element during load transmission through the articular cartilage.

Analytical calculations for simplified joint geometries (Ateshian et al. 1994b; Ateshian and Wang 1995; Kelkar and Ateshian 1995) have shown that the generation of interstitial hydrostatic pressure in the cartilage supports for a period of about 100–200 s more than 90% of the mechanical load, and this has recently been confirmed experimentally (Soltz and Ateshian 1998). The extent of this effect is determined by the incongruity of the joint, the thickness of the cartilage, the properties of the material, and the nature and degree of the loading applied. The authors calculated that in a congruous joint the interstitial hydrostatic pressure in the articular cartilage is maintained for a longer time than with an incongruous "fit." However, these analytical calculations were carried out on idealized congruous and convexly incongruous cylinders and not on concavely incongruous joint surfaces. It must be remembered that with concave incongruity the contact areas extend rapidly towards the center under higher forces, and that the joint acquires a secondarily congruous configuration. We therefore think it is possible that in concavely incongruous joints there is a combined effect of "load distribution" at the joint surface and rapid generation of interstitial hydrostatic pressure with secondary congruity (under loading). No analysis of these relationships, however, can be undertaken on the basis of a linear elastic model of articular cartilage, as this requires a biphasic formulation of the material and its nonlinear contact (Donzelli and Spilker 1993, 1995, 1998). Such investigations are at present under way (see Sect. 5.5.2).

Lubrication of the surfaces must also be taken into account in considering the effects of joint incongruity. Although the mathematical/biomechanical basis of the articular lubrication, which provides an almost frictionless sliding of the joint surfaces, is not fully understood (Hou et al. 1992; Mow et al. 1993; Mow and Ratcliffe 1997), it is conceivable that loading presses fluid out of the superficial layer of the articular cartilage, and that this allows an improved sliding between the joint surfaces. In terms of the stress distribution and stability of the joint, a central joint reaction force shows certain advantages over eccentric loading (Pauwels 1965, 1980; Kummer 1968; Bullough 1981; An et al. 1990). The conversion of telemetric measurements of the in vivo forces acting on the proximal femur (Bergmann et al. 1993) to an acetabular reference system (Witte et al. 1997) has shown that during walking the joint reaction force follows a remarkably constant course relative to the socket. In a congruous joint, however, this would produce a constant contact area and maximal pressure in the center of the articular surface. Assuming concave incongruity, the contact areas change even when the direction of the reaction force remains constant as a function of the force magnitude, and this should improve the lubrication of the cartilage when the joint is subjected to typical cyclical loading. To what extent this phenomenon plays a part in the functioning of normal cartilage and the development of osteoarthrosis must be the subject of future investigations with biphasic models of the cartilage and an appropriate mathematical formulation of the synovial fluid.

Independently of these mechanical considerations, conclusions may be drawn about the metabolic situation of the chondrocytes. Continual displacement of the contact areas during cyclical loading leads to continuous movement of the synovial fluid and, as indicated above, to an increased exchange between the fluid in the articular cartilage and that in the joint cavity. Greenwald and O'Connor (1971) and Afoke et al. (1984b) have suggested that incongruity is of importance for the nutrition of the cartilage tissue. Cell biological studies show that the metabolic stimulation of the chondrocytes is enhanced by moderate (intermittent) loading, whereas it is depressed by static loading (Sah et al. 1989; Urban 1994; Kim et al. 1995). Concave incongruity may therefore have the effect that atrophy of the cartilage tissue in the peripheral regions of the joint is avoided, and that a continuous displacement of the load-bearing areas during cyclical loading increases the synthetic activity of the cells.

5.2.2
Humeroulnar Joint

It can be clearly demonstrated with the anatomically based model of the humeroulnar joint that the incongruity is responsible for bicentric load transmission with maxima in the ventral and dorsal parts of the joint surface. This bicentric pressure distribution is also observed after the central confluence of the contact areas, a finding which extends earlier experimental results (Eckstein et al. 1994b, 1995a) which determined the contact areas but not the contact pressure. With a joint reaction force of about 0.8 times the body weight, pressures of less than 0.5 MPa were calculated in the depth of the joint at all angles of flexion. This is in contrast to a surface pressure of more than 2 MPa in the ventral and dorsal parts of the joint. The more ventral course of the force with small flexion angles (30°) offers a plausible reason why the surface covering the coronoid process is more heavily loaded. With higher angles of flexion (120°), the course of the joint reaction force is, on the other hand, more dorsal, and the surface facing the olecranon is therefore found to experience a higher pressure. This coincides with an experimental determination of the size of the contact areas in the ventral and dorsal regions of the articular surface at various degrees of flexion under simulated extension against resistance (Eckstein et al. 1995a). In the present model a bicentric pressure distribution is observed for all functionally important flexion angles.

This kind of pressure transmission differs significantly from that which Pauwels (1963, 1965, 1980) and An et al. (1990) calculated for the humeroulnar joint. Both assumed in their analysis a congruous "fit" of the joint surfaces. Pauwels (1963, 1965, 1980) determined the long-term stress on the trochlear notch by superimposing the stress diagrams which he had calculated for various angles of flexion. He came to the conclusion that the maximum stress is located in the depth of the notch, a region where in our models there is no pressure at all on the joint surface. This means that the morphological and functional conclusions from these calculations can have only limited validity, since the incongruity of the joint has an extremely important effect on the pressure distribution.

It has occasionally been argued (e.g., Dalstra and Huiskes 1995) that a slight degree of incongruity between the joint components is compensated for by the deformation of the cartilage, and that it can therefore be neglected when calculating the stress distribution in the joint. The present investigation shows, however, that this conclusion is not justified, and that, when designing models for the analysis of load transmission in the humeroulnar joint the exact degree of incongruity must be taken into account if accurate information is to be obtained.

Regarding osteoarthrotic changes in the human elbow it is known that even in old age the humeroulnar joint seldom shows signs of cartilage degeneration, whereas in the humeroradial joint damage appears much earlier (Goodfellow and Bullough 1967; Morrey 1992). Taking into account the above considerations it appears possible that the concave incongruity of the joint makes an important contribution to its prolonged "survival."

5.2.3
Conclusions About Other Joints

The incongruity of the ankle joint was analyzed in a simplified analytical model by Wynarsky and Greenwald (1980). They also determined the effect of the joint fit on the contact behavior and the pressure distribution, and related the geometric configuration to the "relative immunity" of this joint to the development of primary osteoarthrosis.

As noted above, an incongruity of the joint components during walking has been experimentally established for the hip joint (see Sect. 2.1). Whereas Greenwald and O'Connor (1971) attempted to calculate the effect of incongruity on the surface pressure in the lunate surface, incongruity was neglected in nearly all the computer simulations of load transmission in the hip joint (Vasu et al. 1982; Afoke et al. 1982; Brown and Di Gioia 1984; Rapperport et al. 1985; Carter et al. 1987b; Macirowski et al. 1994; Dalstra and Huiskes 1995). This was mostly because of the lack of algorithms for the nonlinear three-dimensional contact of incongruous joint surfaces. Further developments in the FE method, however, will make such calculations possible in the future and may permit important conclusions about the mechanical factors responsible for coxarthrosis. The simulations presented here show that incongruity must certainly not be neglected in this context, and that this applies to both the hip and the shoulder joints.

5.3
The Relationship Between Joint Incongruity, Mechanoadaptive Bone Remodeling and the Morphology of the Subchondral Bone

5.3.1
General Observations

The predictions of the bone remodeling simulations in the idealized model show that the geometric configuration has very important effects not only on the pressure distribution in the joint surface but also on the stress distribution, and therefore the density, of the subchondral bone. Under certain circumstances the bicentric pressure distribution, which is typical for a concavely incongruous joint, precipitates a bicentric distribution of the subchondral density. Concave incongruity therefore offers a possible way of explaining the bicentric density patterns in the humeroulnar joint (Tillmann 1971, 1978; Eckstein et al. 1994b, 1995a), acetabulum (Oberländer et al. 1973; Müller-Gerbl et al. 1993), and trochlea tali (Müller-Gerbl and Putz 1995; Müller-Gerbl 1998). On the other hand, it is also clear from the simulations that conclusions about joint incongruity and/or long-term loading must not be drawn directly from the subchondral density pattern alone. The fit of the joint and the principle direction of the joint reaction force interact in a complex manner. Definite conclusions are therefore only possible when one of these two parameters is additionally known.

If there is a bicentric density pattern in the socket, it is possible that either a concavely incongruous joint is centrally loaded, or else a congruous or convexly incongruous joint is bicentrically loaded. If a monocentric (central) density pattern is present, we may be dealing with a congruous or convexly incongruous joint with a centrally applied load, or else a concavely incongruous joint which is loaded with great force. This means that definite conclusions about the fit of a joint can only be drawn from CT-OAM when all the details of the loading history are known. This is very difficult, however, in the in vivo situation, for which this noninvasive method is particularly suitable. The external forces can be measured, but the muscle forces present a mathematically indeterminate and therefore insoluble problem, since several muscles contribute to the balance of the joint moments. For this reason it seems to us impossible, at the present state of knowledge, to derive accurate information about alterations of joint incongruity with increasing age from changes in the subchondral density (Müller-Gerbl et al. 1991, 1993; Müller-Gerbl and Putz 1995; Müller-Gerbl 1998). On the other hand, if both the subchondral mineralization pattern and the incongruity of the joint can be determined, conclusions about the loading history of the joint can indeed be drawn. In the case of living persons the incongruity cannot be derived directly by CT since this can demonstrate the bone-cartilage interface but not the articular surface. However, it can be achieved with MRI, and new perspectives which this imaging technique may offer are described in Sect. 5.5.1. A combination of

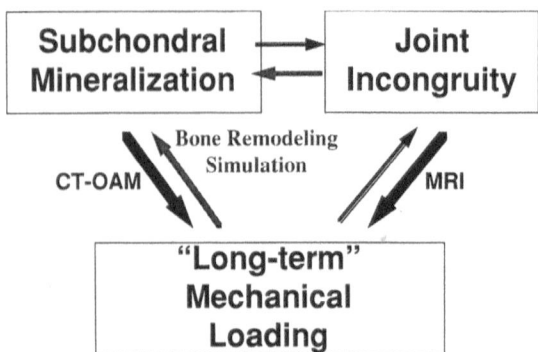

Fig. 47. Interrelationship between subchondral mineralization (density), joint incongruity, and "long-term" mechanical loading

the two techniques may allow the development of methods for determining the average and long-term distribution of the loading of the joint (Fig. 47).

In this connection it is also of interest that a method has been introduced recently by which the long-term loading pattern on a bone can to a certain extent be determined from the density pattern by reversing a bone modeling simulation and by means of an optimization function (Fischer et al. 1993, 1995, 1997; Fischer 1995). The possibilities that this procedure may offer are discussed below (see Sect. 5.5.4).

It seems to us particularly interesting that in incongruous joints the mechanical stresses in the subchondral bone are (depending on the conditions present) distributed differently from those at the articular surface. This is directly opposed to the opinion of Pauwels (1963, 1965, 1980) that the subchondral density represents a "materialized stress diagram" (*verkörpertes Spannungsfeld*) of the pressure acting at the articular surface. In the idealized model the poor correlation between the surface pressure and the predicted subchondral density is due principally to the fact that tensional stresses arise in the latter which, like the compressive stresses, contribute to the functional adaptation of the tissue. Tensional stresses in the subchondral bone of concave joint components have also been suggested by Simkin et al. (1980, 1991) and Dewire and Simkin (1996). It should again be recalled here that the SED is calculated from the stress and strain tensors. It is therefore of no consequence for the prediction of the functional adaptation of the bone whether the SED is due to tensional, compressive, or shear stress. It must also be pointed out that tensional stresses in the subchondral bone are not limited to incongruous joints, they can equally well arise in a congruous joint with peripheral application of the load (by bending the edges of the socket). However, tensional stresses appear in concavely incongruous joints to a quite extraordinary degree, since even with a central load application bending of the socket, and therefore tension of the subchondral and subarticular trabecular bone may occur. The bending can cause tensional stress maxima to appear in places where the joint components are not in contact. If these tensional stresses make an important contribution to the SED in relation to the compressive stresses, the subchondral density pattern reflects the average value of these tangential tensional stresses rather than the normal compressive stresses. To what extent bending of the concavely incongruous sockets actually takes place under physiological conditions depends, in addition to the direction of the applied force, principally upon the individual structure of the joint, i.e., its three-dimensional form, the density of the underlying trabecular bone, the thickness of the cortical bone, and the position of the muscle insertions. An anatomically based model is therefore necessary to answer these questions.

5.3.2
Humeroulnar Joint

No iterative remodeling simulation has been carried out in the anatomically based model, but the relatively uniform distribution of the subchondral SED in the notch suggests that the tissue is well adapted to the mechanical loading. The bone remodeling theory assumes that the density of the tissue is regulated in such a manner that the feedback signal under average loading is the same in all regions. In the comparative model with homogeneous properties of the cortical, trabecular, and subchondral bone there is a bicentric distribution of the SED, and this provides a sound explanation for the bicentric density distribution found in the subchondral bone of the trochlear notch by us and others (Tillmann 1971, 1978). Furthermore, the higher SED in the ulna as compared with the humerus is in agreement with our CT-OAM findings that a higher density is regularly found in the notch. The bicentric density distribution, however, does not agree with the observation by Pauwels (1963, 1965, 1980) that there is a central density maximum in a sagittal saw-cut through an ulna. He explained this by means of superimposed stress diagrams which he calculated assuming congruity of the joint. It is possible that he investigated one of the rare specimens with a continuous joint surface, for which a high degree of congruity has been demonstrated experimentally (Eckstein et al. 1993, 1994b).

Based on the theory of causal histogenesis of Pauwels (1960, 1965, 1980) and Kummer (1963), Tillmann (1971, 1978) suggested that the cartilage-free zone in the depth of the trochlear notch is due to insufficient intermittent loading of the cartilage in that region. Carter et al. (1987a,b, 1991), Carter and Wong (1988) and Wong and Carter (1990) also believed that cartilaginous tissue is preserved only at locations where an adequate hydrostatic pressure occurs regularly. On the other hand, it has been suggested that shear stresses, which were summarized mathematically under the concept of octohedral shear stress, encourage the calcification of the tissue. The authors created a so-called "osteogenic index" from the relationship between hydrostatic pressure and octohedral shear stress which they used for iterative simulations of the advance of the calcification front in joint development. It is easily conceivable that high shear stresses in the depth of the notch – which result from the incongruity – could cause the calcification here to advance as far as the joint surface and bring about an interruption. It should be noted, however, that in the analysis of the differentiation processes in connective tissue the elastic material properties assumed in these studies may only allow a first approximation. More recently, biphasic theories of the differentiation processes in connective tissue have been formulated which also include the effect of mechanically induced fluid flow (Prendergast et al. 1995, 1997b). These are dealt with in more detail below (see Sect. 5.5.5).

The subdivision of the ulnar joint surface, as noted above, is not found in all persons. In 30% of cases the medial, and in 5% both the medial and lateral parts of the joint present a continuous articular surface (Tillmann 1971, 1978). It has been shown experimentally that with continuous surfaces the ventrodorsal contact areas merge in the depth of the trochlear notch under smaller forces than in fully divided joints (Eckstein et al. 1993, 1994b). Our results with CT-OAM have confirmed the finding of Tillmann (1971, 1978) that there is a wider central subchondral density maximum in the joints with a continuous surface. On these grounds it may be assumed that the specimen examined by Pauwels (1963, 1965, 1980) had such a type of joint surface. As noted above, however, it cannot be deduced with certainty from the subchondral density pattern alone whether the corresponding joints are congruous, or whether they have been subjected to more severe loading, although the experimental data presented (Eckstein et al. 1993, 1994b) do indicate a greater degree of congruity. MRI scans of a congruous humeroulnar joint shown in Sect. 2.1 (from Eckstein et al. 1996a) also suggests that this was a specimen with a continuous articular surface.

In the humeroulnar model (central load application at a flexion angle of 90°) the maxima of the subchondral SED are located in a similar position to that of the maximal joint contact pressure. However, the relative distribution of the SED in the socket does not closely reflect the distribution of the contact area pressure in the joint, which can be attributed to the spreading of the socket when the head sinks deeper into the incongruous joint. With other angles of flexion and an oblique action of the load on the ulna, a more marked deviation of the subchondral SED from the contact stress on the joint surface appears which is to be attributed to the higher contribution of tensional stresses to the SED. At a flexion angle of 30° these amount to about four times the compressive stresses in the subchondral bone.

The architecture of the subarticular trabecular bone reflects very clearly the mechanical behavior of the concavely incongruous joint. In particular, comparison with the representation of the principal stress (predominantly tension) in the vector graphs of the idealized model shows a good agreement with the trajectorial construction of the subarticular bone of the trochlear notch. Whereas the (narrow) trabeculae running perpendicular to the joint surface take up the ventral and dorsal pressure, the stout trabeculae in the central region, which lie tangential to the joint surface, seem to demonstrate adaptation to high tensional stresses. This clearly supports the appearance of subchondral tensional stresses in concavely incongruous joints. Anisotropic simulations of bone remodeling (see Sect. 5.5.3) will show whether the trabecular architecture of the ulna can indeed be derived from initially homogeneous, isotropic conditions on the basis of joint incongruity (Jacobs et al. 1998b). An extension of the algorithms so far developed for determining the long-term loading from density distributions (Fischer et al. 1993, 1995, 1997; Fischer 1995; Sect. 5.5.4) could under certain circumstances make additional use of microstructural factors to provide more precise information about the long-term loading on the joint.

A further indication of the occurrence of tensional stress is provided by the sagittal course of the split lines in the tangential (surface) cartilage layer of the ulna (Hultkrantz 1898; Tillmann 1978) and by the sagittal split lines established by us in the subchondral plate of the trochlear notch. It can be assumed that the collagen fibrils of the cartilage and bone orient themselves so as to coincide with the most frequent direction of tension. Therefore, they can be regarded as the morphological correlate of the tensional stresses in the concavely incongruous humeroulnar joint.

5.3.3
Conclusions About Other Joints

Regarding the humeroradial joint, the central location of the density maxima which we found with CT-OAM confirms the prediction of the model for congruous or convexly incongruous joints. Such a geometric configuration has been established both by examination of the contact areas (Bünck 1990) and by MRI (Eckstein et al. 1996a).

The association between bicentric load transmission (Wynarski and Greenwald 1980) and a bicentric mineralization pattern of the trochlea tali (Müller-Gerbl and Putz 1995; Müller-Gerbl 1998) has been discussed above.

In the acetabulum the bicentric distribution of the subchondral mineralization with maxima in the transitional regions from the roof to the anterior horn and from the roof to the posterior horn of the lunate surface appears to agree well with the suggestion that the joint is incongruous and that complete contact in the roof is only reached when the load is high (Bullough et al. 1968, 1973; Greenwald and O'Connor 1971; Greenwald and Haynes 1972; Goodfellow and Mitsou 1977; Afoke et al. 1980; Day et al. 1975; Mizrahi et al. 1981; Brown and Shaw 1983; Miyanaga et al. 1984; Adams and Swanson 1985; Afoke et al. 1987; Eckstein et al. 1997a; von Eisenhart-Rothe et al. 1997, 1998, 1999). More recent investigations, however, in which the subchondral density pattern was compared directly with the location of the contact areas and the contact stress (Fuji Prescale Film) have found no agreement between central contact and a monocentric density pattern or between bicentric contact and a ventrodorsal density pattern (von Eisenhart-Rothe et al. 1997). On the contrary, specimens with a wide joint space at the acetabular roof and contact at the anterior and posterior horns have even shown a very striking central density maximum in the roof. According to our investigations, this maximum can be attributed to tangential tensional stresses in the subchondral plate of the concavely incongruous hip joint which result from the spreading of the socket during load transmission. That the socket of the hip joint does actually tend to spread under normal loading, is supported by the experimental evidence of load-dependent stretching of the transverse acetabular ligament (Löhe et al. 1996). These findings also support our suggestion that potential age-dependent changes in the degree of incongruity cannot be deduced from the distribution of the subchondral mineralization of the acetabulum alone.

In the shoulder, Müller-Gerbl 1998 have reported bicentric, ventrodorsal subchondral density patterns of the glenoid cavity. However, since the glenohumeral joint has been shown to yield either a convex incongruity, a congruous configuration, or superoinferior load bearing areas (Kirsch et al. 1993; Soslowsky et al. 1992b; Conzen and Eckstein 1999), this density pattern cannot be explained by bicentric, ventrodorsal pressure maxima. The ventrodorsal orientation of the subchondral split-lines in the glenoid (Vogt et al. 1999) nevertheless suggest that tension occurs along this plane, most likely from bending during eccentric loading. The glenoid is well supported by bone tissue superiorly and inferiorly but not in its anterior and posterior aspects. This bending of the subchondral bone may thus be responsible for the ventrodorsal density pattern of the subchondral bone rather than a specific articular geometry.

In their article "Roman arches, human joints and disease," Simkin et al. (1980) reported their examination of the structural characteristics of the subarticular trabecular bone of various human joints by microradiography. It was established that the thickness of the subchondral plate of the concave joint component exceeds several times that of the convex side. Indeed, they expressed the opinion that the head of the joint, in a manner similar to a Roman arch, is essentially subjected to pressure, whereas in the concave joint component tensional stresses tangential to the joint surface appear. These observations were supported by morphometric examination of the subchondral bone in the canine shoulder joint (Simkin et al. 1991) and in primates (Dewire and Simkin 1996). However, these authors seem to have overlooked the fact that the joints they examined can show a concave incongruity, and that this is associated with much higher tensional stress than is to be expected in a congruous joint. In view of our suggestion that concave joint incongruity can have both mechanically and biologically a "protective" effect against the development of osteoarthrosis, one wonders whether under these circumstances the title "Gothic arches, human joints and ease" would not have been more suitable.

5.4
Phylogenetic and Ontogenetic Origin of Joint Incongruity

A great deal less is known about the appearance of joint incongruity in phylogeny and ontogeny than about its mechanical and functional significance. Goodfellow and Mitsou (1977) suggested that a joint possesses certain control and regulation mechanisms which are sensitive to pressure, and which by means of differential growth bring about and maintain incongruity during the course of life. In an animal experiment, 6-month-old rabbits with open epiphyses were deprived of loading of the hip on one side, either by direct disarticulation, muscle resection or disarticulation of the ipsilateral knee joint. The animals were sacrificed at various times (1–6 weeks after the operation). The incongruity of the immobilized joint was then compared with that of the contralateral side by a staining method. The authors found ventrodorsal contact areas in the anterior and posterior horns on the nontreated side, but involvement of the acetabular roof with the contact area 1–2 weeks after immobilization. It must be pointed out, however, that the forces in this study were applied manually and were not standardized. Moreover, the staining method only permits the incongruity to be indirectly assessed. The authors also admit that their results can be explained alternatively by a reduction in the mechanical stiffness of the cartilage.

We know of no modern experimental investigations dealing directly with the phylogenetic or ontogenetic origin of joint incongruity. Numerous studies have been reported concerning early joint development. These have established that with immobilization or reduction in the mobility normal morphogenesis is possible up to a stage shortly before cavitation (Fell 1925; Murray 1926; Murray and Selby 1930; Fell and Canti 1934; Hamburger and Waugh 1940), but that no joint cavity develops under these conditions, and that the joint components are flattened (Drachman and Sokoloff 1966; Sullivan 1966; Murray and Drachman 1969; Ruano-Gil et al. 1978; Persson 1983; Kitchener and Laing 1993). In vitro culture of the developing limbs has also shown that without movement the joint cavity does not appear (Pellgrini 1933; Fell and Canti 1934; Bradley 1970), but that with passive movement a joint cavity develops (Lelkes 1958).

Other workers have suggested that mechanical stimulation is of only minor importance for the normal morphogenesis of a joint, and have observed that in transplanted (noninnervated) limbs a cavity forms, at least in some joints (Hamburger 1928, 1939; Hamburger and Waugh 1940). However, they also observed an alteration in the fine modeling of the joint surfaces. Mitrovic (1982) reported that a cavity did initially develop after paralysis with a neuromuscular blocker, but that this was followed later by fusion of the joint components. Investigations by Ruano-Gil et al. (1985) have shown that reserpin-induced hypermobility in chickens is followed by the appearance of an unusually large joint cavity.

The effect of mechanical factors on the growth of developing articular cartilage has been examined in numerous studies. Hueter (1862) and Volkmann (1862) suggested a negative relationship between the magnitude of the forces and the growth of cartilage. On the other hand, McMaster and Weinert (1972) demonstrated both an increased rate of mitosis and the stimulation of cartilage proliferation in the chick as a result of moderate mechanical loading. They also showed an effect of mechanically induced cell displacement on the principal direction of growth. The authors considered that the mechanisms described might be significant for the formation of congruous joints. Recently an increasing number of molecular systems have been discovered which are mechanosensitive, and which affect cell growth (Vandenburgh 1992).

Recent studies have revealed the pivotal role played in initiating tissue separation and joint cavitation by nonadherent components, such as hyaluronic acid (HA) and its interaction with cell-associated HA-binding proteins (HABPs) and HA receptors, such as CD44 (Craig et al. 1990; Edwards et al. 1994; Pitsillides et al. 1995b; Dowthwaite et al. 1997). It has been demonstrated that disruption of HA-HABP interactions prevents functional joint cavity formation (Dowthwaite et al. 1997). Specifically, the in ovo administration of the shorter HA oligosaccharides, which occupy the HABPs, has been demonstrated to alter the local glycosaminoglycan metabolism, to decrease the CD44 and moesin expression (moesin being an actin capping CD44 linking protein), and to result in a failure of joint cavitation (Dowthwaite et al. 1997). Immobilization experiments (Ward and Pitsillides 1998) have established that flaccid paralysis (load and movement removed, induced by pancuronium bromide) and spastic paralysis (removal of movement alone, induced by decamethonium bromide) both inhibit joint cavitation in a similar way before the state of overt element separation, but affect the maintenance of joint cavities differently after their separation. Immobilization leads to a selective decrease in the activity of UDPGD (a key enzyme in the synthesis of HA), a decrease in nonsulfated HA and a decrease in CD44 and moesin expression in the presumptive joint line of immobilized limbs (Ward and Pitsillides 1998). These and other results suggest that, prior to cavitation, movement exerts a dominant effect on the differentiation of cells bordering the presumptive joint spaces, and that after cavitation has been fully established, physiological loading contributes substantially to maintaining functionally competent diathrodial joints (Ward et al. 1999). Ward and Pitsillides (1999) have recently demonstrated that chick p42 phosphorylated mitogen-activated protein kinase/extracellular-signal-regulated kinase (MAP/ERK; an enzyme which is also involved in the intracellular events by which endothelial cells respond to mechanical stimuli) shows a clear joint line-selective distribution during normal joint cavitation. Immunolabeling for constitutively active ERK-1, however, was not evident at the joint line in limbs which had been paralyzed.

Dowthwaite et al. (1999) observed that in cells isolated from developing joints of stage 42 chick embryos uniaxial four-point-bending at 3800 microstrain significantly increased HA synthesis after 6 h of mechanical stimulation, compared with static and flow samples. At 24 h, UDPGD activity, CD44 labeling, and the total HA binding sites were increased significantly in the mechanically stimulated samples, the changes being associated with alterations in the phosphorylation of CD44 and moesin. These observations suggest that complex signaling events control both CD44 expression and HA synthesis, and that synthesis of HA and binding to HA are closely coordinated. Another experiment in cells isolated from the developing articular surface of embryonic chick joints showed that activated ERK-1 expression was increased to similar extents 24 h after a 10 min application of either dynamic or static strain (Ward and Pitsillides, 1999), but that fluid flow caused a more dramatic increase in ERK-1 expression. Dynamic and static strain (but not flow) increased the UDPGD activity by around 50%, and the HA concentrations by 65%–80% in the medium (Ward and Pitsillides 1999). These results indicate discrepant effects of distinct components of the mechanical environment. Although some important principles are beginning to emerge, it is still unknown by what mechanism mechanical stimuli control or affect joint development and maintenance, or cause the joints to become physiologically incongruous.

Embryonic movements are known to start in the chick during the fourth day of development and reach a maximum between days 11 and 13 (Bekoff 1981). Interestingly, this coincides with the period at which immobilzation in ovo is most fatal (Ward and Pitsillides, 1998). During this period, an average of more than 20 movements per minute has been recorded. These movements are not uncoordinated but follow an organized pattern (Watson and Bekoff 1990). In human embryos distinguishable movements begin in the 6th to 7th weeks of intrauterine life, and from the 8th week onwards rapid twitching movements (so-called "startles") are observed, as well as generalized and more long-continued movements of the limbs, body, and head (De Vries et al. 1982; Gnirs and Schneider 1994). Isolated movements of the limbs occur from the 9th week of pregnancy. From about the start of the second trimester (14th week) the fetus shows the same pattern of movement as at the time of birth. A maximum of activity is reached in the 34th week (Gnirs and Schneider 1994), when oscillating movements appear to take place, mostly at a frequency of about one cycle per minute (Robertson et al. 1982).

It is thus evident that, during the critical period of joint development and growth, the limbs are already moving actively. Because of the short lever arms of the muscles the hydrodynamic resistance of the surrounding fluid to the limbs and the firm resistance of the uterine wall, the mechanical loading of the joints should be considerable even at this stage. The above observations indicate that the genetically determined developmental and growth processes are effectively modulated by mechanical factors, and that a complex interaction of intrinsic (genetic and hormonal) and extrinsic (mechanical) events are involved in regulating the ontogenetic development of synovial joints. As noted above, however, is remains unclear how and when human joints become physiologically incongruous, or to what degree the generation of incongruity is controlled by mechanical stimuli.

Because the reparative processes appearing during joint degeneration generally recapitulate embryonic developmental and regulatory mechanisms (Archer 1994), a sound understanding of the underlying ontogenetic mechanobiological processes is a prerequisite for finding an effective treatment for joint disease. Further investigations in cell biology and animal experiments are needed to clarify the degree to which genetic and mechanical factors affect these developmental processes. The large number of new transgenic animal models, new possibilities of analyzing cell biological processes in intact tissue, and more subtle techniques for regulating the mechanical environment of single cells and tissues suggest that decisive progress will be made in this area within the next few years. Such experiments could make essential contributions to achieving a more refined theory of joint development, joint function, and joint degeneration. These experiments should be extended by the further development of high-resolution noninvasive imaging techniques, which are suitable for the precise determination of joint shape and the quantitative distribution of the tissues in living persons or experimental animals. In addition to this, advances in the mathematical formulation and computer simulation of the mechanobiological interaction of the differentiation, growth, and regulation of the connective tissues are required to support the experimental and morphological investigations. Only the combination of these techniques can lead to a complete understanding of the relationship of form and function, and of the diseases of human joints. Together these could make it possible to design more effective therapeutic treatments than those presently available. Current investigations and future perspectives in this field are presented in the following section.

5.5
Perspectives and Further Investigations

5.5.1
Noninvasive, Quantitative Determination of Joint Incongruity and Cartilage Thickness by MRI

The purpose of the present investigation has been to answer the major questions about pressure transmission and adaptive processes in the subchondral bone of incongruous joints. For the modeling it has therefore been possible to resort to invasive techniques for determining the incongruity, contact areas, and cartilage thickness, which offer certain advantages because of their high spatial resolution. However, predicting the effect of an operation (a correction osteotomy, for instance) on the joint pressure of a living person requires specific individual models and noninvasive methods which permit realistic modeling without interfering with the health of the subject. As noted above, the quantitative distribution of the bone tissue and, within certain limits, its material properties can be determined noninvasively with CT. However, because of the lack of image contrast between the articular cartilage and the joint cavity, neither the incongruity nor the cartilage thickness can be delineated by this method.

New possibilities have become available with MRI. This technique requires no ionizing radiation, and delineates both the interface between the joint cavity and the articular cartilage and that between cartilage and subchondral bone with high contrast if suitable pulse sequences are used. Recent work has shown that with high resolution, T1-weighted fat-suppressed gradient echo sequences, both the volume (Peterfy et al. 1994, 1995; Eckstein et al. 1994c, 1995c, 1998a–c; Marshall et al. 1995; Sittek et al. 1996) and thickness of the cartilage (Eckstein et al. 1995c, 1996b, 1997b, 1998a,b; Sittek et al. 1996) can be accurately and reproducibly measured. Digital image analysis techniques have been developed which make it possible to obtain a three-dimensional analysis of cartilage thickness, independently of the position and orientation of the original section plane (Lösch et al. 1995, 1997; Münsterer et al. 1996; Eckstein et al. 1996c; Haubner et al. 1997; Stammberger et al. 1999). The articular cartilage is first reconstructed in three dimensions from serial images, and then the minimal distance between the articular surface and the bone-cartilage interface is calculated for each imaging point by custom-made software. The possible resolution of this procedure at the present time is for the knee joint $1.5 \times 0.31 \times 0.31$ mm^3. In the elbow a resolution of $1 \times 0.2 \times 0.2$ mm^3 has recently been achieved by applying a gradient echo sequence with selective water excitation (Springer et al. 1998; Figs. 48, 49). This technique opens new possibilities for developing specific individual models of the joints of particular patients, and these could, for example, provide for better preoperative planning in the fields of orthopedic or traumatic surgery.

5.5.2
Biphasic Modeling of the Articular Cartilage

The distribution of the load between the fluid and solid (the proteoglycan-collagen matrix) phases is of decisive importance for the mechanical efficiency of the cartilage. Ateshian et al. (1994b), Ateshian and Wang (1995), and Kelkar and Ateshian (1995) have shown analytically for idealized joint shapes (contact between convex cylinders of various radii) that with short-term loading the greater part of the mechanical load is supported by the interstitial hydrostatic pressure rather than by the matrix. However, the calculation of the distribution of the load between the two phases for real joint shapes requires a numerical analysis by means of a computer model. A further prerequisite for this is the formulation of a biphasic material model of the articular cartilage as well as adequate algorithms for characterizing the nonlinear contact behavior of two biphasic cartilage layers under loading. Such formulations have recently been developed (Donzelli and Spilker 1993, 1995, 1998). In cooperation with this group we have therefore started to carry out simulations of load transmission through physiologically incongruous joints, based on our quantitative measurements of the joint space width in the elbow. Based on the geometry of the idealized models presented in this investigation, calculations have been made for various forms and degrees of incongruity under short-term loading of 35 N (1-s loading duration; Donzelli et al. 1997). In this way it has been possible to calculate for a given time-point the interstitial hydrostatic pressure (p), the so-called elastic stress (given by the deformation of the matrix) and the so-called solid stress (obtained from the elastic stress and a hydrostatic pressure component). The ratio obtained for the interstitial hydrostatic pressure and the elastic or solid stress provides a measure for that part of the load which is taken up by fluid, where negative values attribute a greater significance to this "protective" mechanism.

It has been found that hydrostatic pressurization occurs in all types of joints, with the given combination (load 35 N, $t=1$ s), the highest ratio between the interstitial hydrostatic pressure and the elastic (matrix) stress appearing in the congruous joint. The load distribution in concavely incongruous joints is more favorable than in those with convex incongruity. It must be remembered, however, that this simulation has been dealing with very small loads because higher forces bring about a deformation

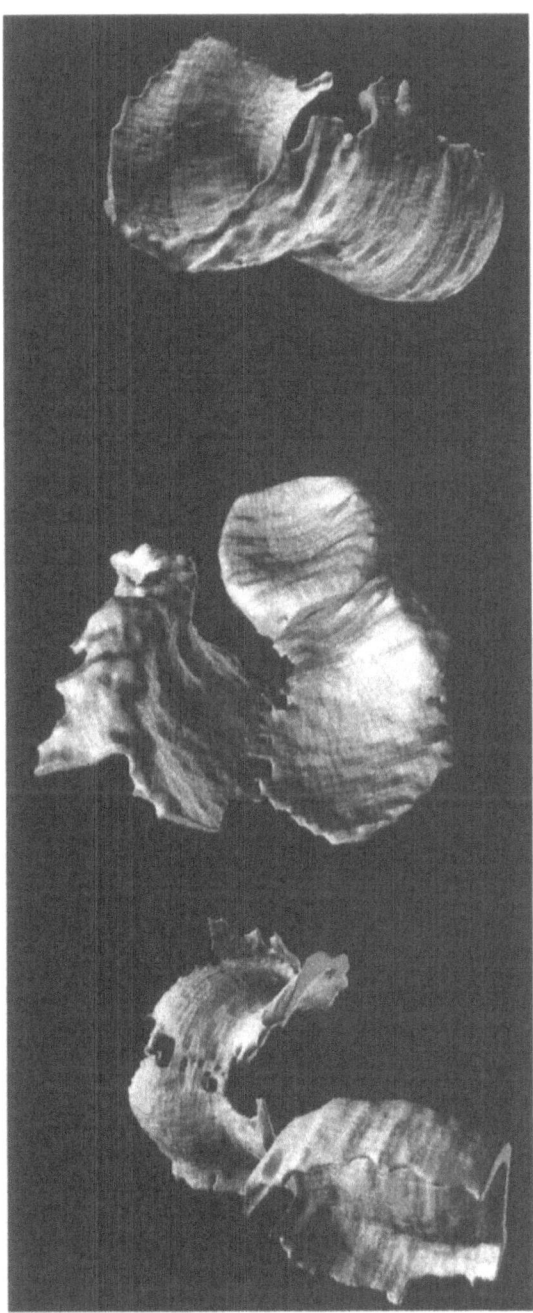

Fig. 48. Three-dimensional reconstruction of the elbow joint cartilages from high-resolution MRI (3D gradient echo sequence with selective water excitation, spatial resolution $1 \times 0.25 \times 0.25$ mm^3). With permission from Springer V, Graichen H, Stammberger T, Englmeier K-H, Reiser M, Eckstein F (1998) Nichtinvasive Analyse des Knorpelvolumens und der Knorpeldicke im menschlichen Ellbogengelenk mittels MRT. Ann Anat 180:331—338 (Copyright Gustav Fischer 1998)

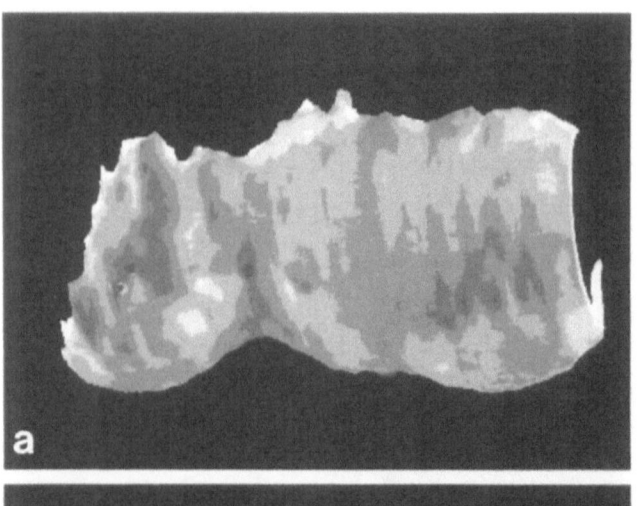

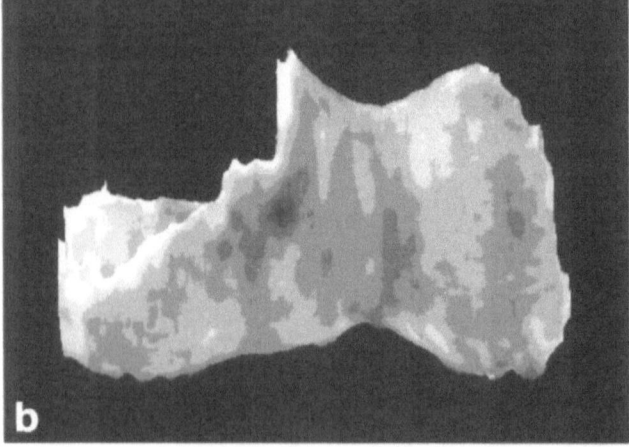

Fig. 49 a,b. Thickness of the humeral joint cartilages computed from the 3D reconstruction shown in Fig. 48. With permission from Springer V, Graichen H, Stammberger T, Englmeier K-H, Reiser M, Eckstein F (1998) Nichtinvasive Analyse des Knorpelvolumens und der Knorpeldicke im menschlichen Ellbogengelenk mittels MRT. Ann Anat 180:331—338 (Copyright Gustav Fischer 1998)

of the cartilage which cannot be adequately treated with existing algorithms. Since the introduction of greater loads in concavely incongruous joints causes a deformation which secondarily produces congruity, it is conceivable that under these conditions the relationships are altered and that there is a more favorable load distribution than in a congruous joint because of the more important contribution of the joint periphery.

Further development of biphasic material models and nonlinear contact algorithms should make it possible in the near future to carry out simulations of the load transmission in incongruous joints under higher joint reaction forces. This should provide exact information about the extent to which the proteoglycan-collagen matrix is actually mechanically strained in incongruous joints, and therefore make it possible to draw more specific conclusions about the role of incongruity in the etiology of degenerative joint diseases.

5.5.3
Anisotropic Simulation of Bone Remodeling

As noted above, bone tissue has an anisotropic structure, i.e., its material properties are to a considerable extent dependent upon the direction of the applied load. Fyhrie and Carter (1986) have introduced a mathematical theory by which not only the density but also the direction of the trabeculae is optimized relative to the mechanical loading (see below).

Based on the theory that the osteocytes act as sensors of mechanical loads, and that the osteoblasts and osteoclasts (which regulate bone density) are under the control of the osteocytes, Mullender et al. (1994) developed a model in which the "sensors" and "actors" are spatially separated. In this model the local mechanical stimulus is not only determined by the local stresses and strains but also by the distance from the surrounding sensors. They introduced a distance parameter D (defining the range of effect of each osteocyte) and obtained discontinuous "checkerboard" patterns (Weinans et al. 1992) from initially homogeneous conditions for small values of D (sensors and actors at the same site), but stable trabecular-like structures for higher values of D (overlapping ranges of effects for the sensors). Mullender and Huiskes (1995) have shown that the thickness, number, and separation of the trabeculae in the model depend directly on the magnitude of D, and that the trabecular-like structures align with the principal stress direction. The osteocytes were shown to be the more likely candidates for providing a mechanically efficient structure than the bone lining cells (Mullender and Huiskes 1997). Van Rietbergen et al. (1996) and Mullender et al. (1998) generated plate-like and rod-like structures in small-scale three-dimensional analyses, corresponding to the morphology of the real trabecular network, and these also aligned with the principal stress direction.

The technique of large-scale FE modeling for entire bones (van Rietbergen et al. 1999; Ulrich et al. 1999, also see below) may eventually lead to remodeling simulations in which the load-adaptive behavior of individual trabeculae can be treated directly. However, a method of accounting for the directional mechanical properties of trabecluar bone in remodeling simulations at the continuum level is nonetheless valuable and is the direction taken in several recent bone remodeling theories. Fernandes et al. (1999) pursued an approach inspired by the original trajectorial theory of Wolff (1892). In this formulation trabecular bone is allowed to take on orthotropic apparent material properties, i.e., properties that depend on direction but are limited to trabecular structures in which trabeculae intersect at right angles. The investigators have developed the theoretical conditions and computer algorithms needed to find the optimal distribution of bone apparent density and orientation for each element in a given FE model. One interesting finding was that they were not able to find a single orientation of the orthotropic model which was appropriate for several load cases simultaneously. From this, one may conclude that the trajectorial condition of Wolff (1892) (specifically the assertion that trabeculae always intersect at right angles) does not hold under multiple loading conditions. This is in agreement with other studies that have suggested that trabeculae may align nonorthogonally (in contradiction with

Wolff's law), in order to reduce the shear-coupling occurring under multidirectional loading near joint surfaces (Hert 1992, Pidaparti and Turner 1997).

Jacobs et al. (1997) took an alternative theoretical approach in which no assumptions were made about limitations to the apparent material properties due to the microstructural organization of trabecular bone. They developed and implemented a theory that predicted optimal material properties which could depend on direction and could predict a response to multiple load cases simultaneously. Although the predictions for the proximal femur (Jacobs et al. 1997) agreed qualitatively with the typical orientation of the trabeculae and quantitatively with measurements of the orientation-dependent bone stiffness (Reilly and Burstein 1975; van Buskirk and Ashmann 1981), a concern with this approach is that it is possible for the computer to predict apparent material properties that do not correspond to any feasible trabecular microstructure. Recently a novel approach has been proposed which seems an appropriate compromise (Rodrigues et al. 1999). With this method FE simulations are carried out at two different length scales. First, a global model is constructed for the bone or joint of interest. Next, a microstructural model is employed at each element to determine the optimal trabecular morphology for that point in the global model. This information is then used to compute the apparent material properties for the global model. Although promising, this approach is very new and has yet to be validated in a quantitative fashion.

5.5.4
Computer-Aided Determination of the Long-Term Loading from the Subchondral Density

Fischer et al. (1993, 1995) and Fischer (1995) developed a computer-aided method based on an inversion of the bone remodeling theory described by Beaupre et al. (1990a,b) and certain optimizing criteria. With this, they have been able to select load cases that are able to produce a density pattern that resembles as closely as possible the density distribution in a given structure. Using this method it is possible to draw objective conclusions about the dominant loading pattern during normal daily use (Fischer 1995), for instance, from CT-derived density distributions.

In a recent study of hip joints with a normal CCD angle, a large CCD angle (coxa valga), and a small CCD angle (coxa vara), a density-based analysis revealed a larger angle of the joint reaction force with the sagittal plane than the static analyses by Amtmann and Kummer (1968) and Pauwels (1973) for the stance-phase of walking (Eckstein et al. 1996d; Fischer et al. 1997, 1999). With regard to the relative change in the dominant loading with coxa valga and vara, however, there was a good agreement with the calculated values. It should be noted that the density-based analysis accounts for the integral of all mechanical loads and not for a single event. It is therefore suitable for analyzing the long-term loading on the bone, although it has so far only been possible to reliably calculate the main direction and relative magnitudes and not the absolute magnitudes of the joint reaction force (Fischer 1995; Fischer et al. 1997, 1999). It should also be noted that the results depend somewhat on assumptions in the underlying bone remodeling theory, but whereas these can affect the absolute magnitudes, they should not greatly affect the relative load magnitudes (patterns).

Since the subchondral density patterns of incongruous joints react very sensitively to the load magnitude, however, it is conceivable that the introduction of the (incongruous) cartilage layer and the joint contact into the density-based analysis of joint loading would produce improved results that can also provide reasonable estimates of the load magnitudes. This nevertheless requires both the simultaneous solution of the nonlinear contact of two cartilage layers and the optimization function developed by Fischer (1995) and Fischer et al. (1993, 1995). The implementation and testing of these on idealized models of incongruous joints is anticipated. A calculation based on the inversion of anisotropic algorithms could also make use of microstructural parameters of the subchondral and subarticular trabecular bone to obtain better predictions of the long-term loading from morphological parameters.

5.5.5
Computer Simulation of Growth
and Differentiation of the Connective Tissues

As discussed in Sect. 5.4., the growth and differentiation processes of the connective tissues play a key role in the morphogenesis of joints, although it remains unclear to what extent these follow an established genetic program or are modulated by mechanical factors. According to Roux (1912), tensional stress leads to the production of dense fibrous tissue, shear stress to that of cartilage and compressive stress to that of bone. Pauwels (1960) criticized this idea and suggested that strains caused by the distortion of shape of the tissue, but not by changes in its volume, are the specific stimulus for the production of collagen fibrils or dense fibrous tissue. On the other hand, Pauwels (1960) regarded volume changes in the tissue as the specific stimulus for the production of cartilage, which is preserved as long as a certain critical value is not exceeded and pressure is supplemented by intermittent shear. If, however, the tissue is surrounded by an inflexible shell, in his opinion, a swelling of the cells brings about a continuous increase in the hydrostatic pressure which causes the cartilage to be replaced by bone.

A formal mathematical relationship between the differentiation of the connective tissues and their mechanical environment has been developed by Carter et al. (1987a,c). Their model predicts that ossification of the tissue is accelerated by intermittent shear stresses and inhibited by hydrostatic pressure. In FE analyses a so-called "osteogenic index" has been used to describe the main events in the morphogenesis of the human femur (Carter et al. 1987a), the sternum (Wong and Carter 1988), and the process of epiphyseal endochondral ossification (Carter and Wong 1988; Wong and Carter 1990). It has been pointed out that, contrary to the findings of Pauwels, the ossification center presents a region of high shear stresses and low hydrostatic pressure, and that the hydrostatic pressure in the area of the future articular cartilage is very high. It should be noted, however, that in this analysis the connective tissues have been regarded as elastic, and this excludes the possible effect of fluid flow on cell differentiation.

Using the experimental data of Soballe (1993), Prendergast et al. (1995, 1997b) performed a biphasic FE analysis which treated the fluid and solid phases of the tissue separately, and predicted effects on the tissue differentiation at bone-implant interfaces. They showed that the biophysical stimuli acting on the connective tissues

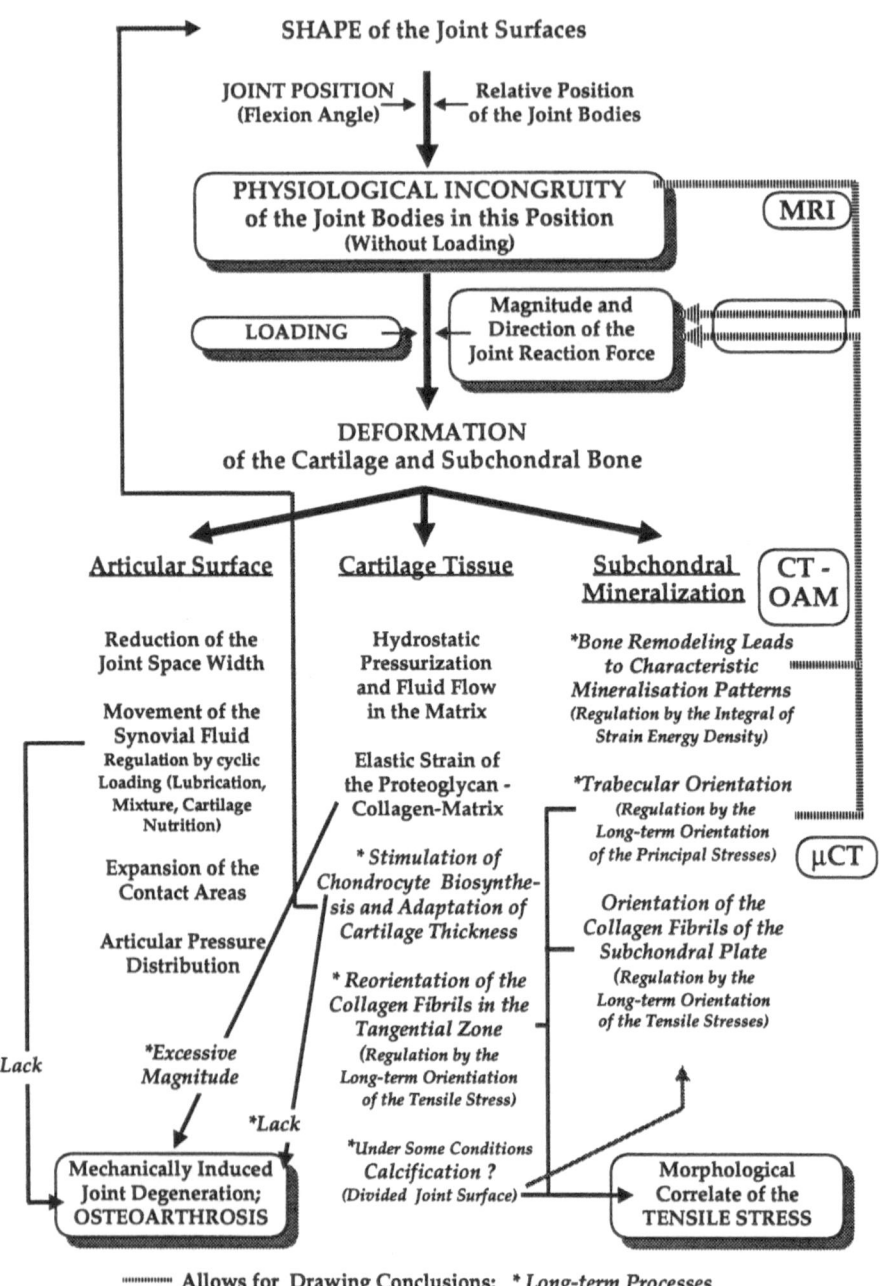

SHAPE of the Joint Surfaces

JOINT POSITION (Flexion Angle) → ← Relative Position of the Joint Bodies

PHYSIOLOGICAL INCONGRUITY of the Joint Bodies in this Position (Without Loading)   MRI

LOADING → ← Magnitude and Direction of the Joint Reaction Force

DEFORMATION of the Cartilage and Subchondral Bone

Articular Surface | Cartilage Tissue | Subchondral Mineralization   CT - OAM

Articular Surface

Reduction of the Joint Space Width

Movement of the Synovial Fluid *Regulation by cyclic Loading (Lubrication, Mixture, Cartilage Nutrition)*

Expansion of the Contact Areas

Articular Pressure Distribution

*Lack

*Excessive Magnitude

*Lack

Cartilage Tissue

Hydrostatic Pressurization and Fluid Flow in the Matrix

Elastic Strain of the Proteoglycan - Collagen-Matrix

* Stimulation of Chondrocyte Biosynthesis and Adaptation of Cartilage Thickness

* Reorientation of the Collagen Fibrils in the Tangential Zone *(Regulation by the Long-term Orientiation of the Tensile Stress)*

Under Some Conditions Calcification ? (Divided Joint Surface)

Subchondral Mineralization

Bone Remodeling Leads to Characteristic Mineralisation Patterns (Regulation by the Integral of Strain Energy Density)

Trabecular Orientation (Regulation by the Long-term Orientation of the Principal Stresses)   μCT

Orientation of the Collagen Fibrils of the Subchondral Plate *(Regulation by the Long-term Orientation of the Tensile Stresses)*

Mechanically Induced Joint Degeneration; **OSTEOARTHROSIS**

Morphological Correlate of the **TENSILE STRESS**

▪▪▪▪▪▪ Allows for Drawing Conclusions;   * Long-term Processes

Fig. 50. Flow chart showing hypothetical consequences of joint incongruity on load transmission, cartilage strain, subchondral bone remodeling, and the etiology of osteoarthrosis

change dramatically during the course of the experiment. To begin with, the mechanical environment is characterized by low cyclic pressure, rapid fluid flow, and a high shear strain. Under these circumstances the fibroblasts appear to produce a collagen matrix which stiffens the solid phase and reduces the permeability of the tissue. The resultant increase in hydrostatic pressure and decrease in fluid flow are mechanical conditions under which the chondrocytes are thought to produce fibrocartilage. As a consequence of this change in extracellular matrix synthesis, the mechanical conditions become stable, and the fluid flow and shear strain in the tissue decrease considerably. This in turn encourages the proliferation of osteoblasts and eventually ossification. The mechanobiological theory developed here (Prendergast et al. 1995, 1996, 1997b) can, for instance, be used in a FE model to analyze why in the humeroulnar joint calcification at the depth of the trochlear notch advances up to the joint surface, and why a divided articular surface is to be found in most adults. In this way the general validity of the theory presented could be tested and the mechanical basis of this specific phenomenon determined.

Concerning the growth processes of the connective tissue, van der Meulen et al. (1993, 1995) distinguished between a "basic biological rate of growth" and a "mechanobiological component," the latter being regulated by mechanical stresses. Based on computer models, in which the development of the cross-sectional form of tubular bones was investigated, they calculated the effect of the mechanobiological component on the apposition and resorption processes at the periosteal and endosteal surfaces. Changes in the diaphyseal structure of long tubular bones, which depend on growth and age, were successfully predicted and revealed a high degree of dependence on mechanical stimulation. Based on this theory, Stevens et al. (1996, 1999) developed models of the epiphyses to predict the endochondral ossification processes and the development of an articular cartilage layer on the basis of a mechanobiological component and also a biological growth function.

Brodland and Clausi (1994) employed a large strain-finite element formulation to describe how contractile forces generate shape changes during morphogenesis. Heegaard et al. (1995, 1999a) presented a FE model of an embryonic finger joint. Accounting for the theories of Fick (1890) on the significance of the position of the tendon insertion for the resulting joint shape, and assuming that tensional stresses accelerate the growth rate (whereas compressive stresses reduce it), they predicted the development of a concave joint component from a convex anlage by simulating complete flexion-extension cycles. Heegaard (1998, 1999b) recently used such a mechanically driven mathematical growth model to study the stability of the pattern formation process during joint morphogenesis and how mechanical forces may affect joint congruence. He defined an intrinsic biological baseline stimulus, which he assumed to be proportional to the local chondrocyte density. The effect of mechanobiological factors on growth was modeled by a stimulus whose exact nature remains unknown but is taken to depend on the osteogenic index (Carter and Wong, 1988). Compressive pressure was assumed to lead to a local decrease in the surface radius of curvature but a state of tensile stress to a local increase. A generic diarthrodial joint model was used to compute the kinematics resulting from muscle contractions, and the stress distribution that was postulated to modulate the cartilage anlagen growth. When including only the biological baseline growth in this morphogenesis model, the articular segments grew into noncongruent shapes, and the path leading to the final shape was unstable. Small variations in the biological stimulus resulted in large variations in the final joint shapes, but including the mechanobiological stimulus stabilized the joint morphogenesis process and growth converged to a common pattern. Force appeared to be the most important parameter in guiding the shape of the joint to a converged state. Applying a constant load but starting from various initial configurations eventually produced the same shapes. Starting from a given shape but simulating the chondrogenesis process with various loads, on the other hand, produced different converged shapes. The authors described conditions under which the joint became concavely (bicentric) and convexly incongruous (monocentric). Starting with a nearly congruent joint but with a slight degree of convex incongruity, and applying small or moderate loads during morphogenesis, the articulation eventually converged to a convexly incongruous joint. However, with higher load magnitude the degree of congruence increased (the radii of curvature becoming identical) until a critical stage was reached. Past this point, the articulation developed into a concavely incongruous joint. These findings support the hypothesis that epigenetic factors, such as mechanical stress, stabilize and guide the underlying biological growth and differentiation processes.

Its is evident that these theories and models themselves are still at an "embryonic" stage. As noted above, further experiments in cell biology and cell mechanics are necessary to complement and validate these fragmentary and still speculative theories. However, using various examples of tissue regeneration and embryogenesis, FE models will make it possible to design more searching and focused experiments, and to test such mechanobiological theories. It is to be hoped that the parallel development of experimental and mathematical approaches will lead to a conclusive theory about the development, functional adaptation, regeneration and degeneration of the connective tissues, and that this will permit deeper understanding of the interaction of mechanical factors and biological processes involved in both the embryonic establishment and the subsequent maintenance, adaptation, and degeneration of synovial joints.

6 Conclusions

In answering the questions posed in the introduction, the following conclusions may be reached:

Incongruity and Pressure Distribution. The natural incongruity of synovial joints has a profound effect on the pressure distribution throughout the articular surface. The incongruity is not completely compensated by the deformation of the articular cartilage, and the resulting stress distribution therefore does not correspond to that of a congruous joint. Concave incongruity (deeper joint socket) produces an inhomogeneous distribution of the contact stress, with maxima at the periphery and low pressure at the center. As the magnitude of the load increases, the contact areas widen in the direction of the center of the joint surface; nevertheless, an inhomogeneous pressure distribution with peripheral maxima is maintained.

Incongruity and Subchondral Bone Remodeling. In the bone remodeling simulations performed here various types of incongruity produced different subchondral density patterns. Concave incongruity leads to a typical bicentric distribution, but with increased loading central monocentric mineralization patterns were also observed. By using radiological methods (CT-OAM), corresponding mineralization patterns were demonstrated in joints with different types of "fit." The results of bone remodeling simulations are in agreement with the morphological findings in joints with corresponding shape.

Pressure Distribution and Subchondral Bone Remodeling. In incongruous joints there is no striking agreement between the long-term distribution of the contact stresses on the articular surface and the distribution of the subchondral bone density. This can be attributed to the fact that the stress distribution in the subchondral bone deviates considerably from that at the joint surface, owing to spreading of the joint socket. Thus in contrast to earlier theories, the subchondral bone density does not offer a "materialized stress diagram" of the pressure on the articular surface.

The Value of CT Osteoabsorptiometry. Both the incongruity and the loads acting on the joint determine the local stress on the articular surface and the subchondral bone. These two factors interact in a complex manner and can therefore be predicted by CT-OAM only if one of these two parameters is exactly known. Since reliable data on the long-term mechanical loading in joints of living persons is virtually unobtainable, it does not seem possible with the present state of knowledge to draw conclusions from the subchondral density, either about the shape of the joint or about life-time changes in incongruity/congruity. If, on the other hand, the incongruity could be analyzed, by means of MRI, for example, objective deductions about the loading history of the joint could become possible.

Incongruity and Subchondral Tension. Particularly in concavely incongruous joints, tensional stresses play a significant role in the loading and functional adaptation of the subchondral bone. These make an essential contribution to those mechanical signals that regulate the local bone density. During central load application the subchondral tensional stresses in the trochlear notch are of a similar magnitude as the compressive stresses, and they can even exceed the latter several times during eccentric loading. Whereas the compressive stresses exhibit a bicentric distribution in the humeroulnar joint, the tensional stress maxima are located more centrally in the ulnar joint surface. This is mainly why the stress distribution in the subchondral bone deviates from that on the joint surface.

Morphological Correlates of Subchondral Tension. The architecture of the subarticular trabecular bone and the sagittal arrangement of collagen fibrils in the subchondral plate of the trochlear notch both are an expression of the adaptation to tensional stress that occurs during pressure transmission in the concave components of incongruous joints. They can therefore be regarded as corresponding morphologically to the subchondral tension predicted by the computer models, and they also demonstrate that the tensional stresses are large enough to cause a specific microstructural adaptation of the bone tissue.

Our study shows that anatomically based models of synovial joints are required to perform appropriate analyses of the pressure transmission and the functional adaptive processes in the connective tissues. The normal incongruity of joints must be considered in remodeling simulations if realistic data are to be obtained. It can be shown that computer simulation represents a method of choice by which – on the level of highly complex nonlinear systems – important mechanisms and principles can be revealed with regard to morphological structures, physiological functions, and mechanobiological regulatory processes that cannot be achieved by other techniques. Predictions made by the simulations could be confirmed by the experimental and morphological findings.

The results of the current study suggest that the concave incongruity is part of a basic structural principle which helps to optimize the pressure transmission in the joint and, at the same time, ensures that the joint cartilage is well lubricated and nourished. The ontogenetic and phylogenetic development of joint incongruity has not yet been elucidated, but recent experimental data indicate that mechanical stimuli play a crucial role in this process. Further studies in cell biology and cell mechanics as well as animal experimentation are required to propose a more refined theory of mechanobiological processes, such as differentiation, growth, adaptation and degeneration in the connective tissues in general, and in the actual organ, "the joint," in particular. These will have to be supplemented by the further development of high-resolution, noninvasive imaging techniques and digital postprocessing as well as by advances in the application of mathematical formulation and computer simulation of mechanobiological processes. A combination of molecular and cell biological techniques, together with noninvasive imaging procedures and new computer-simulation methods, can lead to a better understanding of the form/function relationship in human joints. Such an improved understanding ought to facilitate the application of more specific and effective therapeutic approaches to joint diseases in the future than is possible at the present time.

References

Adam C, Eckstein F, Milz S, Putz R (1998) The distribution of cartilage thickness within the joints of the lower limb of elderly individuals. J Anat 193:203–214

Adams D, Swanson SA (1985) Direct measurement of local pressures in the cadaveric human hip joint during simulated level walking. Ann Rheum Dis 44:658–666

Afoke NY, Byers PD, Hutton WC (1980) The incongruous hip joint. A casting study. J Bone Joint Surg Br 62:511–514

Afoke NYW, Byers PD, Hutton WC (1982) A finite element study of the human hip joint. Eng Med 11:17–24

Afoke NY, Byers PD, Hutton WC (1984a) The incongruous hip joint: a loading study. Ann Rheum Dis 43:295–301

Afoke A, Hutton WC, Byers PD (1984b) Synovial fluid circulation in the hip joint. Med Hypotheses 15:81–86

Afoke NY, Byers PD, Hutton WC (1987) Contact pressures in the human hip joint. J Bone Joint Surg Br 69:536–541

Akizuki S, Mow VC, Muller F, Pita JC, Howell DS, Manicourt DH (1986) Tensile properties of human knee joint cartilage. I. Influence of ionic conditions, weight bearing, and fibrillation on the tensile modulus. J Orthop Res 4:379–392

Amtmann E, Kummer B (1968) Die Beanspruchung des menschlichen Hüftgelenks. II. Größe und Richtung der Hüftgelenksresultierenden in der Frontalebene. Z Anat Entw Gesch 127:286–314

An KN, Kwak BM, Chao EY, Morrey BF (1984) Determination of muscle and joint forces: a new technique to solve the indeterminate problem. J Biomech Eng 106:364–367

An KN, Himeno S, Tsumura H, Kawai T, Chao EY (1990) Pressure distribution on articular surfaces: application to joint stability evaluation. J Biomech 23:1013–1020

Archer CW (1994) Skeletal development and osteoarthritis. Ann Rheum Dis 53:624–630

Ascenzi A (1993) Biomechanics and Galileo Galilei. J Biomech 26:95 100

Ateshian GA, Soslowsky LJ, Mow VC (1991) Quantitation of articular surface topography and cartilage thickness in knee joints using stereophotogrammetry. J Biomech 24:761–776

Ateshian GA, Kwak SD, Soslowsky LJ, Mow VC (1994a) A stereophotogrammetric method for determining in situ contact areas in diarthrodial joints, and a comparison with other methods. J Biomech 27:111–124

Ateshian GA, Lai WM, Zhu WB, Mow VC (1994b) An asymptotic solution for the contact of two biphasic cartilage layers. J Biomech 27:347–1360

Ateshian GA, Ark JW, Rosenwasser MP, Pawluk RJ, Soslowsky LJ, Mow VC (1995a) Contact areas in the thumb carpometacarpal joint. J Orthop Res 13:450–458

Ateshian GA, Wang HA (1995b) theoretical solution for the frictionless rolling contact of cylindrical biphasic articular cartilage layers. J Biomech 28:1341–1355

Ateshian GA (1997) A theoretical formulation for boundary friction in articular cartilage. J Biomech Eng 119:81–86

Athanasiou KA, Rosenwasser MP, Buckwalter JA, Malinin TI, Mow VC (1991) Interspecies comparisons of in situ intrinsic mechanical properties of distal femoral cartilage. J Orthop Res 9:330–340

Bachrach NM, Valhmu WB, Stazzone E, Ratcliffe A, Lai WM, Mow VC (1995) Changes in proteoglycan synthesis of chondrocytes in articular cartilage are associated with the time dependent changes in their mechanical environment. J Biomech 28:1561–1569

Badley EM, Thompson RP, Wood PH (1978) The prevalence and severity of major disabling conditions – a reappraisal of the government social survey on the handicapped and impaired in Great Britain. Int J Epidemiol 7:145–151

Beaupre GS, Hayes WC (1985) Finite element analysis of a three-dimensional open-celled model for trabecular bone. J Biomech Eng 107:249–256

Beaupre GS, Orr TE, Carter DR (1990a) An approach for time dependent bone modeling and remodeling theoretical development. J Orthop Res. 8:651–661

Beaupre GS, Orr TE, Carter DR (1990b) An approach for time dependent bone modeling and remodeling application: a preliminary remodeling simulation. J Orthop Res 8:662–670

Bekoff A (1981) Embryonic development of chick motor behaviour. Trends Neurosci:181–184

Benninghoff A (1931) Über Leitsysteme der Knochenkompakta. Gegenbaurs Morphol Jahrb 65:11–44

Bergmann G, Graichen F, Rohlmann A (1993) Hip joint loading during walking and running, measured in two patients. J Biomech 26:969–990

Bertram JE, Swartz SM (1991) The "law of bone transformation": a case of crying Wolff? Biol Rev Cambridge Philos Soc 66:245–273

Bourgery DIM (1832) Traite complet de l'anatomie de l'homme. I. Osteologie (Paris)

Bradley SJ (1970) An analysis of self differentiation of chick limb buds in chorio-allantoic grafts. J Anat 107:479–490

Brand R, Mehr D, Pardubsky P, Buckwakter J, Martin M (1997) Does tenascin help adapt cells to mechanical stress? Trans Eur Orthop Res Soc 7:26

Brand RA, Pedersen DR, Davy DT, Kotzar GM, Heiple KG, Goldberg VM (1994) Comparison of hip force calculations and measurements in the same patient. J Arthroplasty 9:45–51

Brighton CT, Strafford B, Gross SB, Leatherwood DF, Williams JL, Pollack SR (1991) The proliferative and synthetic response of isolated calvarial bone cells of rats to cyclic biaxial mechanical strain. J Bone Joint Surg Am 73:320–331

Brighton CT, Sennett BJ, Farmer JC, Iannotti JP, Hansen CA, Williams JL, Williamson J (1992) The inositol phosphate pathway as a mediator in the proliferative response of rat calvarial bone cells to cyclical biaxial mechanical strain. J Orthop Res 10:385–393

Brodland GW, Clausi DA (1994) Embryonic tissue morphogenesis modelled by FEM. J Biomech Eng 116:146–155

Brown TD, DiGioia AM (1984) A contact coupled finite element analysis of the natural adult hip. J Biomech 17:437–448

Brown TD, Ferguson AB Jr (1980) Mechanical property distributions in the cancellous bone of the human proximal femur. Acta Orthop Scand 51:429–437

Brown TD, Shaw DT (1983) In vitro contact stress distributions in the natural human hip. J Biomech 16:373–384

Bünck S (1990) Krümmungs und Kontaktflächenverhältnisse der Articulatio humeroradialis. Anat Anz 171:45–53

Bullough PG (1981) The geometry of diarthrodial joints, its physiologic maintenance, and the possible significance of age related changes in geometry to load distribution and the development of osteoarthritis. Clin Orthop 156:61–66

Bullough PG, Jagannath A (1993) The morphology of the calcification front in articular cartilage. Its significance in joint function. J Bone Joint Surg Br 65:72–78

Bullough P, Goodfellow J, Greenwald AS, O'Connor J (1968) Incongruent surfaces in the human hip joint. Nature 217:1290

Bullough P, Goodfellow J, O'Conner J (1973) The relationship between degenerative changes and load bearing in the human hip. J Bone Joint Surg Br 55:746–758

Burger EH, Klein Nulend J, Semeins CM, Ajubi NE, Nijweide PJ (1996) Osteocytes but not periostal fibroblasts produce nitric oxide (NO) in response to pulsatile fluid flow. Trans Orthop Res Soc 21:31

Cabe GH, Hong JH, Hipp JA, Snyder BD (1998) Structural rigidity predicts the load capacity of trabecular bone with simulated osteolytic defects. Trans Orthop Res Soc 2:547

Carter DR, Hayes WC, Schurman DJ (1976) Fatigue life of compact bone II. Effects of microstructure and density. J Biomech 9:211–218

Carter DR, Hayes WC (1977) The compressive behavior of bone as a two phase porous structure. J Bone Joint Surg Am 59:954–962

Carter DR, Schwab GH, Spengler DM (1980) Tensile fracture of cancellous bone. Acta Orthop Scand 51:733–741

Carter DR (1984) Mechanical loading histories and cortical bone remodeling. Calcif Tissue Int 36 [Suppl 1]:S19–S24

Carter DR (1987) Mechanical loading history and skeletal biology. J Biomech 20:1095–1109

Carter DR, Orr TE, Fyhrie DP, Schurman DJ (1987a) Influences of mechanical stress on prenatal and postnatal skeletal development. Clin Orthop 219:237–250

Carter DR, Rapperport DJ, Fyhrie DP, Schurman DJ (1987b) Relation of coxarthrosis to stresses and morphogenesis. A finite element analysis. Acta Orthop Scand 58:611–619

Carter DR, Fyhrie DP, Whalen RT (1987c) Trabecular bone density and loading history: regulation of connective tissue biology by mechanical energy. J Biomech 20:785–794

Carter DR, Wong M (1988) The role of mechanical loading histories in the development of diarthrodial joints. J Orthop Res 6:804–816

Carter DR, Orr TE, Fyhrie DP (1989) Relationships between loading history and femoral cancellous bone architecture. J Biomech 22:231–244

Carter DR, Wong M, Orr TE (1991) Musculoskeletal ontogeny, phylogeny, and functional adaptation. J Biomech 24 Suppl 1:3–16

Carvalho RS, Scott JE, Yen EH (1995) The effects of mechanical stimulation on the distribution of beta 1 integrin and expression of beta 1 integrin mRNA in TE 85 human osteosarcoma cells. Arch Oral Biol 40:257–264

Caterson D, Lowther DA (1978) Changes in the metabolism of the proteoglycans from sheep articular cartilage in response to mechanical stress. Biochim Biophys Acta 540:412–422

Choi K, Kuhn JL, Ciarelli MJ, Goldstein SA (1990) The elastic moduli of human subchondral, trabecular, and cortical bone tissue and the size dependency of cortical bone modulus. J Biomech 23:1103–1013

Clift SE (1992) Finite element analysis in cartilage biomechanics. J Biomed Eng 14:217–221

Clover J, Dodds RA, Gowen M (1992) Integrin subunit expression by human osteoblasts and osteoclasts in situ and in culture. J Cell Sci 103:267–271

Conzen A, Eckstein F, Landgraf J, Anetzberger H, Putz R (1997) Normal joint pressure and load-bearing areas in the human shoulder as a function of joint position and loading. Trans Eur Orthop Res Soc 7:163

Conzen A, Eckstein F (1999) Quantitative determination of articular pressure in the human shoulder joint. J Shoulder Elbow Surg (in press)

Cowin SC (1984) Mechanical modeling of the stress adaptation process in bone. Calcif Tissue Int 36 [Suppl 1]:S98–S103

Cowin SC, Hegedus DH (1976) Bone remodeling. I. A theory of adaptive elasticity. J Elasticity 6:323–326

Cowin SC, Nachlinger RR (1976) Bone remodeling. III. Uniqueness and stability in adaptive elasticity theory. J Elasticity 8:285–295

Cowin SC, Van Buskirk WC (1978) Internal bone remodeling induced by a medullary pin. J Biomech 11:269–275

Cowin SC, Van Buskirk WC (1979) Surface bone remodeling induced by a medullary pin. J Biomech 12:269–276

Cowin SC, Moss Salentijn L, Moss ML (1991) Candidates for the mechanosensory system in bone. J Biomech Eng 113:191–197

Cowin SC, Weinbaum S, Zeng Y (1995) A case for bone canaliculi as the anatomical site of strain generated potentials. J Biomech 28:1281–1297

Craig FM, Bayliss MT, Bentley G, Archer CW (1990) A role for hyaluronan in joint development. J Anat 171:17–23

Currey JD (1995) The validation of algorithms used to explain adaptive remodeling in bone. In: Odgaard A, Weinans H (eds) Bone structure and remodeling, recent advances in human biology, vol 2. World Scientific, Singapore, pp 9–14

Dalstra M, Huiskes R (1995) Load transfer across the pelvic bone. J Biomech 28:715–724

Darwin C (1872) The origin of species, 6th edn. New American Library, New York

Day WH, Swanson SA, Freeman MA (1975) Contact pressures in the loaded human cadaver hip. J Bone Joint Surg Br 57:302–313

De Vries JIP, Visser GHA, Prechtl HFR (1982) The emergence of fetal behaviour. I. Qualitative aspects. Early Hum Dev 7:301–322

Dewire P, Simkin PA (1996) Subchondral plate thickness reflects tensile stress in the primate acetabulum. J Orthop Res 14:838–841

Diaz A, Sigmund A (1995) Checkerboard patterns in layout optimization. Struct Optimization 10:40–45

Donkers MJ, An KN, Chao EY, Morrey BF (1993) Hand position affects elbow joint load during push up exercise. J Biomech 26:625–632

Donzelli PS, Spilker RL (1993) A finite element formulation for contact of biphasic materials: evaluation for plane problems. In: Tarbell J (ed) Advances in bioengineering. American Society of Medical Engineers, New York, pp 47–50

Donzelli PS, Spilker RL (1995) An iterative contact detection algorithm for a mixed penalty biphasic finite element. American Society of Medical Engineers Bioengineering Conference BED 29:169–170

Donzelli P, Spilker RLA (1998) A contact finite element formulation for biological soft hydrated tissues. Comput Methods Appl Mech Eng 153:68–79

Donzelli P, Eckstein F, Putz R, Spilker RL (1997) Physiological joint incongruity significantly affects the load partitioning between the solid and fluid phases of articular cartilage. Trans Orthop Res Soc 22:82

Doty SB (1981) Morphological evidence of gap junctions between bone cells. Calcif Tissue Int 33:509–512

Dowthwaite GP, Edwards JCW, Pitsillides AA (1997) The role of hyaluronan binding proteins during joint development. Trans Orthop Res Soc 22:59

Dowthwaite GP, Flannery CR, Pitsillides AA, Archer CW (1999) The role of mechanical strain in the formation of diarthrodial joints: a possible mechanism controlling hyaluronan production. Trans Orthop Res Soc 23:12

Drachman DB, Sokoloff L (1966) The role of movement in embryonic joint development. Dev Biol 14:401–420

Duncan RL, Turner CH (1995) Mechanotransduction and the functional response of bone to mechanical strain. Calcif Tissue Int 57:344–358

Eckstein F, Müller Gerbl M, Putz R (1992) Distribution of subchondral bone density and cartilage thickness in the human patella. J Anat 180:425–433

Eckstein F, Löhe F, Schulte E, Müller Gerbl M, Milz S, Putz R (1993) Physiological incongruity of the humero ulnar joint: a functional principle of optimized stress distribution acting upon articulating surfaces? Anat Embryol 188:449–455

Eckstein F, Merz B, Schmid P, Putz R (1994a) The influence of geometry on the stress distribution in joints a finite element analysis. Anat Embryol 189:545–552

Eckstein F, Löhe F, Müller Gerbl M, Steinlechner M, Putz R (1994b) Stress distribution in the trochlear notch. A model of bicentric load transmission through joints. J Bone Joint Surg Br 76:647–653

Eckstein F, Sittek H, Milz S, Putz R, Reiser M (1994c) The morphology of articular cartilage assessed by magnetic resonance imaging (MRI). Reproducibility and anatomical correlation. Surg Radiol Anat 16:429–438

Eckstein F, Löhe F, Hillebrand S, Bergmann M, Schulte E, Milz S, Putz R (1995a) Morphomechanics of the humero ulnar joint. I. Joint space width and contact areas as a function of load and flexion angle. Anat Rec 243:318–326

Eckstein F, Merz B, Müller Gerbl M, Holzknecht N, Pleier M, Putz R (1995b) Morphomechanics of the humero ulnar joint. II. Concave incongruity determines the distribution of load and subchondral mineralization. Anat Rec 243:327–335

Eckstein F, Sittek H, Milz S, Schulte E, Kiefer B, Reiser M, Putz R (1995c) The potential of magnetic resonance imaging (MRI) for quantifying articular cartilage thickness a methodological study. Clin Biomech 10:434–440

Eckstein F, Merz B, Sittek H, Kolem H, Reiser M, Putz R (1996a) Geometry to pressure relationship in the human elbow joint: a qualitative analysis using MRI and finite elements. Eur J Anat 1:23–30

Eckstein F, Sittek H, Gavazzeni A, Schulte E, Milz S, Kiefer B, Reiser M, Putz R (1996b) Magnetic resonance chondro-crassometry (MR CCM): a method for accurate determination of articular cartilage thickness? Magn Reson Med 35:89–96

Eckstein F, Gavazzeni A, Sittek H, Haubner M, Lösch A, Milz S, Englmeier KH, Schulte E, Putz R, Reiser M (1996c) Determination of knee joint cartilage thickness using three dimensional magnetic resonance chondro-crassometry (3D-MR-CCM). Magn Reson Med 36:256–265

Eckstein F, Fischer K, Becker C, Putz R (1996d) Dichte-basierte mathematische Analyse der längerfristigen Hüftgelenksbeanspruchung aus CT Daten. Ann Anat 179 [Suppl]:245

Eckstein F, von Eisenhart Rothe R, Landgraf J, Adam C, Löhe F, Müller Gerbl M, Putz R (1997a) Quantitative analysis of incongruity, contact areas and cartilage thickness in the human hip joint. Acta Anat 158:192–204

Eckstein F, Adam C, Sittek H, Becker C, Milz S, Schulte E, Reiser M, Putz R (1997b) Non-invasive determination of topographical cartilage thickness maps using magnetic resonance imaging (MRI) optimization and comparison with other techniques. J Biomech 30:285–289

Eckstein F, Schnier M, Haubner, Priebsch J, Glaser C, Englmeier K-H, Reiser M (1998a) Accuracy of three-dimensional knee joint cartilage volume and thickness measurements with MRI. Clin Orthop 352:137–148

Eckstein F, Westhoff J, Sittek H, Maag K-P, Haubner M, Faber S, Englmeier K-H, Reiser M (1998b) In vivo reproducibility of three-dimensional cartilage volume and thickness measurements with magnetic resonance imaging. AJR Am J Roentgenol 170:593–597

Eckstein F, Winzheimer M, Westhoff J, Schnier M, Haubner M, Englmeier K-H, Reiser M, Putz R (1998c) Quantitative relationships of normal cartilage volumes of the human knee joint – assessment by magnetic resonance imaging. Anat Embryol 197:383–390

Edwards JC, Wilkinson LS, Jones HM, Soothill P, Henderson KJ, Worrall JG, Pitsillides AA (1994) The formation of human synovial joint cavities: a possible role for hyaluronan and CD44 in altered interzone cohesion. J Anat 185:355–367

von Eisenhart-Rothe R, Eckstein F, Müller-Gerbl M, Landgraf J, Rock C, Putz R (1997) Direct comparison of contact areas, contact stress and subchondral mineralization in human hip-joint specimens. Anat Embryol 195:279–288

Eisenhart-Rothe von R, Witte H, Steinlechner M, Müller-Gerbl M, Putz R, Eckstein F (1998) Quantification of joint incongruity and contact pressure distribution in the human hip based on in vivo measurements of joint forces. Trans Orthop Res Soc 2:839

Eisenhart-Rothe von R, Adam C, Steinlechner M, Müller-Gerbl M, Eckstein F (1999) Quantitative determination of joint incongruity and pressure distribution during simulated gait, and cartilage thickness in the human hip joint. J Orthop Res 17:532–539

Engh CA, Bobyn JD (1984) Biologic fixation of a modified Moore prosthesis. Part II. Evaluation of adaptive femoral bone modeling. Hip (unnumbered):110–132

Feldkamp LA Goldstein SA, Parfitt AM, Jesion G, Kleerekoper M (1989) The direct examination of three-dimensional bone architecture in vitro by computed tomography. J Bone Miner Res 4:3–11

Fell HB (1925) The histogenesis of cartilage and bone in the long bones of the embryonic fowl. J Morphol Physiol 40:417–459

Fell HB, Canti RC (1934) Experiments on the development in vitro of the avian knee joint. Proc R Soc 116:316–351

Felson DT (1988) Epidemiology of hip and knee osteoarthritis. Epidemiol Prev 10:1–28

Felson DT (1990) Osteoarthritis. Rheum Dis Clin North Am 16:499–512

Fernandes P, Rodrigues H, Jacobs CA (1999) Model of bone adaptation using a global optimisation criterion based on the trajectorial theory of Wolff. Comput Methods Biomech Biomed Eng (in press)

Fick R (1890) Über die Form der Gelenkflächen. Arch Anat Physiol Anat Abt [Suppl]:391–402

Firoozbakhsh K, Cowin SC (1981) An analytical model of Pauwels' functional adaptation mechanism in bone. J Biomech Eng 103:246–252

Fischer KJ (1995) Correspondence between bone density distributions and mechanical loading. Thesis, Stanford University

Fischer KJ, Jacobs CR, Carter DR (1993) Determination of bone and joint loads from bone density distributions. Trans Orthop Res Soc 18:529

Fischer KJ, Jacobs CR, Carter DR (1995) Computational method for determination of bone and joint loads using bone density distributions. J Biomech 28:1127–1135

Fischer KJ, Eckstein F, Becker C, Jacobs CR (1997) Calculation of femoral loading histories for coxa vara and coxa valga from bone density distributions. Trans Orthop Res Soc 22:889

Fischer KJ, Eckstein F, Becker C (1999) Density-based load estimation predicts altered femoral load directions for coxa vara and coxa valga. J Muskuloskeletal Res 3:83–92

Frost HM (1964) The laws of of bone structure. Thomas, Springfield

Frost HM (1983) Bone histomorphometry: analysis of trabecular bone dynamics. In: Recker RR (ed) Bone histomorphometry: techniques and interpretation. CRC, Boca Raton, pp 109–131

Frost HM (1986) Intermediary organization of the skeleton. CRC, Boca Raton

Frost HM (1987) The mechanostat: a proposed pathogenic mechanism of osteoporoses and the bone mass effects of mechanical and nonmechanical agents. Bone Miner 2:73–85

Fukubayashi T, Kurosawa H (1980) The contact area and pressure distribution pattern of the knee. A study of normal and osteoarthrotic knee joints. Acta Orthop Scand 51:871–879

Fung YC (1993) Biomechanics mechanical properties of living tissues, 2nd edn. Springer, Berlin Heidelberg New York

Fyhrie DP, Carter DR (1986) A unifying principle relating stress to trabecular bone morphology. J Orthop Res 4:304–317

Fyhrie DP, Carter DR (1990) Femoral head apparent density distribution predicted from bone stresses. J Biomech 23:1–10

Gailit J, Pierschbacher M, Clark RA (1993) Expression of functional alpha 4 beta 1 integrin by human dermal fibroblasts. J Invest Dermatol 100:323–328

Gibson LJ (1985) The mechanical behaviour of cancellous bone. J Biomech 18:317–328

Gnirs J, Schneider KT (1994) Fetale Verhaltenszustände und Bewegungsaktivität. Gynäkologe 27:136–145

Goldsmith AA, Hayes A, Clift SE (1996) Application of finite elements to the stress analysis of articular cartilage. Med Eng Phys 18:89–98

Goodfellow JW, Bullough PG (1967) The pattern of ageing of the articular cartilage of the elbow joint. J Bone Joint Surg Br 49:175–181

Goodfellow J, Mitsou A (1977) Joint surface incongruity and its maintenance: an experimental study. J Bone Joint Surg Br 59 B:446–451

Graichen H, Lochmüller E-M, Wolf E, Langkabel R, Stammberger T, Haubner M, Renner-Müller I, Englmeier K-H, Eckstein F (1999) A non-destructive technique for 3D microstructural phenotypic characterization of bones in genetically altered mice – preliminary data in growth hormone transgenic animals and normal controls. Anat Embryol 199:239–248

Greenwald AS (1991) Biomechanics of the hip. In: Steinberg ME (ed) The hip and its disorders. Saunders, Philadelphia, pp 47–56

Greenwald AS, Haynes DW (1972) Weight bearing areas in the human hip joint. J Bone Joint Surg Br 54:157–163

Greenwald AS, O'Connor JJ (1971) The transmission of load through the human hip joint. J Biomech 4:507–528

Grodzinski AJ, Berger E, Hung HK, Frank EH, Hunziker EB (1999) Compression of cartilage alters the morphology of intracellular organelles: a potential link between mechanical stimulation and aggrecan structure. Trans Orthop Res Soc 23:671

Guilak F, (1995) Compression induced changes in the shape and volume of the chondrocyte nucleus. J Biomech 28:1529–1541

Guilak F, Ratcliffe A, Mow VC (1995) Chondrocyte deformation and local tissue strain in articular cartilage: a confocal microscopy study. J Orthop Res 13:410–421

Hackenbroch M (1923) Die Arthrosis deformans der Hüfte. Thieme, Leipzig

Halls AA, Travill R (1964) Transmission of pressures across the elbow joint. Anat Rec 150:243–247

Hamburger V (1928) Die Entwicklung experimentell erzeugter nervenloser und schwach innervierter Extremitäten von Anuren. Roux Arch Entw Mech 114:272

Hamburger V (1938) The development and innervation of transplanted limb primordia of chick embryos. J Exp Zool 80:347–389

Hamburger V, Waugh M (1940) The primary development of the skeleton in nerveless and poorly innervated limb transplants of chick embryos. Physiol Zool 13:367–384

Harrigan TP, Hamilton JJ (1992) An analytical and numerical study of the stability of bone remodelling theories: dependence on microstructural stimulus. J Biomech 25:477–488

Harrigan TP, Hamilton JJ (1993) Bone strain sensation via transmembrane potential changes in surface osteoblasts: loading rate and microstructural implications. J Biomech 26:183–200

Harrigan TP, Jasty M, Mann RW, Harris WH (1988) Limitations of the continuum assumption in cancellous bone. J Biomech 21:269–275

Hart RT, Davy DT, Heiple KG (1984a) A computational method for stress analysis of adaptive elastic materials with a view toward applications in strain induced bone remodeling. J Biomech Eng 106:342–350

Hart RT, Davy DT, Heiple KG (1984b) Mathematical modeling and numerical solutions for functionally dependent bone remodeling. Calcif Tissue Int 36 [Suppl 1]:S104–S109

Harter LV, Hruska KA, Duncan RL (1995) Human osteoblast like cells respond to mechanical strain with increased bone matrix protein production independent of hormonal regulation. Endocrinology 136:528–535

Haubner M, Eckstein F, Lösch A, Schnier M, Sittek H, Becker C, Kolem H, Reiser M, Englmeier KH (1997) A non invasive technique for 3 dimensional assessment of articular cartilage thickness based on MRI Part II Validation with CT arthrography. Magn Reson Imaging 15:805–813

Hayes WC, Snyder BD (1981) Toward a quantitative formulation of Wolff's law in trabecular bone. In: Cowin SC (ed) Mechanical properties of bone. American Society of Medical Engineers, New York, pp 43–68

Hayes WC, Keer LM, Herrmann G, Mockros LF (1972) A mathematical analysis for indentation tests of articular cartilage. J Biomech 5:541–551

Hayes WC, Snyder B, Levine BM, Ramaswamy S (1982) Stress morphology relationship in trabecular bone of the patella, In: Gallagher RH, Simon BR, Johnson PC, Gross JF (eds) Finite element in biomechanics. Wiley, New York, pp 223–269

Heegaard JH, Carter DR, Beaupre GS (1995) A mathematical model for simulating mechanically modulated growth in developing diarthrodial joints. Trans Orthop Res Soc 20:73

Heegaard JH (1998) Stability of joint morphogenesis from a dynamical perspective. J Biomech 31 [Suppl]:128

Heegaard JH, Beaupre GS, Carter DR (1999a) Mechanically modulated cartilage growth may regulate joint surface morphogenesis. J Orthop Res 17:509–517

Heegaard JH (1999b) A mathematical model to control joint congruence during morphogenesis. Trans Orthop Res Soc 24:233

Hegedus DH, Cowin SC (1976) Bone remodeling. II. Small strain adaptive elasticity. J Elasticity 6:337–352

Hehne HJ, (1990) Biomechanics of the patellofemoral joint and its clinical relevance. Clin Orthop 258:73–85

Helminen HJ, Kiviranta I, Säämänen AM, Jurvelin JS, Arokoski J, Oettmeier R, Abendroth K, Roth AJ, Tammi MI (1992) Effect of motion and load on articular cartilage in animal models. In: Kuettner KE, Schleyerbach R, Peyron JG, Hascall VC (eds) Articular cartilage and osteoarthritis. Raven, New York, pp 501–510

Hert J (1992) A new explanation of the cancellous bone architecture. Funct Dev Morph 2:17–24

Hoch DH, Grodzinsky AJ, Koob TJ, Albert ML, Eyre DR (1983) Early changes in material properties of rabbit articular cartilage after meniscectomy. J Orthop Res 1:4–12

Hodge WA, Carlson KL, Fijan RS, Burgess RG, Riley PO, Harris WH, Mann RW (1989) Contact pressures from an instrumented hip endoprosthesis. J Bone Joint Surg Am 71:1378–1386

Hodgskinson R, Currey JD (1990) The effect of variation in structure on the Young's modulus of cancellous bone: a comparison of human and non human material. Proc Inst Mech Eng H 204:115–1121

Hollister SJ, Fyhrie DP, Jepsen KJ, Goldstein SA (1991) Application of homogenisation theory to the study of trabecular bone mechanics. J Biomechanics 24:825–839

Hou JS, Mow VC, Lai WM, Holmes MH (1992) An analysis of the squeeze film lubrication mechanism for articular cartilage. J Biomech 25:247–259

Hueter C (1862) Anatomische Studien an den Extremitätengelenken Neugeborener und Erwachsener. Virchows Arch 25:575

Huiskes R (1995) The law of adaptive bone remodeling: a case for crying Newton? In: Odgaard A, Weinans H (eds) Bone structure and remodeling. Recent advances in human biology, vol 2. World Scientific, Singapore, pp 15–24

Huiskes R, Chao EY (1983) A survey of finite element analysis in orthopedic biomechanics: the first decade. J Biomech 16:385–409

Huiskes R, Hollister SJ (1993) From structure to process, from organ to cell: recent developments of FE analysis in orthopaedic biomechanics. J Biomech Eng 115B:520–527

Huiskes R, Weinans H, Grootenboer HJ, Dalstra M, Fudala B, Slooff TJ (1987) Adaptive bone remodeling theory applied to prosthetic design analysis. J Biomech 20:1135–1150

Hultkranz W (1898) Über die Spaltrichtung der Gelenkknorpel. Verh Anat Ges 12:248–256

Hung CT, Pollack SR, Reilly TM, Brighton CT (1995) Real-time calcium response of cultured bone cells to fluid flow. Clin Orthop 313:256–269

Hynes RO (1992) Integrins: versatility, modulation, and signalling in cell adhesion. Cell 69:11–25

Iannotti JP, Gabriel JP, Schneck SL, Evans BG, Misra S (1992) The normal glenohumeral relationships. An anatomical study of one hundred and forty shoulders. J Bone Joint Surg Am 74:491–500

Jacobs CR (1994) Numerical simulation of bone adaptation to mechanical loading. Thesis, Stanford University, USA

Jacobs CR, Levenston ME, Beaupre GS, Simo JC, Carter DR (1995) Numerical instabilities in bone remodeling simulations: the advantages of a node based finite element approach. J Biomech 28:449–459

Jacobs CR, Simo JC, Beaupr GS, and Carter DR (1997) Adaptive bone remodeling incorporating simultaneous density and anisotropy considerations, J. Biomech 30:603–613

Jacobs CR, Yellowley CE, Davis BR, Zhou Z, Cimbala JM, Donahue HJ (1998a) Differential effect of steady versus oscillating flow on bone cells. J Biomech 31:969–976

Jacobs CR, Eckstein F, Merz B (1998b) Bicentric subchondral bone density patterns and trabecular orientation in incongruous joints are due to bending in the subchondral bone layer. Trans Orthop Res Soc 23:546

Jeffery AK, Blunn GW, Archer CW, Bentley G (1991) Three dimensional collagen architecture in bovine articular cartilage. J Bone Joint Surg Br 73:795–801

Jurvelin J, Kiviranta I, Tammi M, Helminen JH (1986) Softening of canine articular cartilage after immobilization of the knee joint. Clin Orthop 207:246–252

Jurvelin J, Saamanen AM, Arokoski J, Helminen HJ, Kiviranta I, Tammi M (1988) Biomechanical properties of the canine knee articular cartilage as related to matrix proteoglycans and collagen. Eng Med 17:157–162

Kelkar R, Ateshian GA (1995) Contact creep response between a rigid impermeable cylinder and a biphasic cartilage layer using integral transform. American Society of Medical Engineers Bioengineering Conference BED 29:313–314

Kempson GE, Spivey CJ, Swanson SA, Freeman MA (1971) Patterns of cartilage stiffness on normal and degenerate human femoral heads. J Biomech 4:597–609

Kim YJ, Bonassar LJ, Grodzinsky AJ (1997) The role of cartilage streaming potential, fluid flow and pressure in the stimulation of chondrocyte biosynthesis during dynamic compression. J Biomech 28:1055–1066

Kirsch S, Nägerl H, Kubein Meesenburg D (1993) Kinematics and statics of the human shoulder. Abstracts of the International Society of Biomechanics XIVth Congress, Paris 692

Kitchener PD, Laing NG (1993) Brachially innervated ectopic hindlimbs in the chick embryo. I. Limb motility and motor system anatomy during the development of embryonic behavior. J Neurobiol 24:280–299

Kiviranta I, Jurvelin J, Tammi M, Saamanen AM, Helminen HJ (1987) Weight bearing controls glycosaminoglycan concentration and articular cartilage thickness in the knee joints of young beagle dogs. Arthritis Rheum 30:801–809

Kiviranta I, Tammi M, Jurvelin J, Arokoski J, Saamanen AM, Helminen HJ (1994) Articular cartilage thickness and glycosaminoglycan distribution in the young canine knee joint after remobilization of the immobilized limb. J Orthop Res 12:161–167

Klein-Nulend J, Van der Plas A, Semeins CM, Ajubi NE, Frangos JA, Nijweide PJ, Burger EH (1995a) Sensitivity of osteocytes to biomechanical stress in vitro. FASEB J 9:441–445

Klein-Nulend J, Roelofsen J, Sterck JG, Semeins CM, Burger EH (1995b) Mechanical loading stimulates the release of transforming growth factor beta activity by cultured mouse calvariae and periosteal cells. J Cell Physiol 163:115–119

Klein-Nulend J, Burger EH, Semeins CM, Reisz LG, Pilbeam CC (1997) Pulsating fluid flow stimulates prostaglandin release and inducible prostaglandin G/H synthase mRNA Expression in primary mouse bone cells. J Bone Miner Res 12:45–51

Klein-Nulend J, Helfrich MH, MacPherson H, Ralston SH, Joldersma M, Sterck JGH, Semeins CM, Burger EH (1998) Mechanotransduction in bone-involvement of nitric oxine as a rapid cellular messenger molecule. Trans Eur Orthop Res Soc 8:19

Knight MM, Lee DA, Bolton JF, Bader DL (1999) Cell and nucleus deformation in compressed chondrocyte-agarose constructs – implications for mechanotransduction. Trans Orthop Res Soc 23:710

Knothe-Tate ML, Knothe U, Niederer P (1998) Experimental elucidation of mechanical load-induced fluid flow and its potential role in bone metabolism and functional adaptation. Am J Med Sci 316:189–195

Knothe-Tate ML, Niederer P, Knote U (1998) In vivo tracer transport through the lacunocanalicular system of rat bone in an environment devoid of mechanical loading. Bone 22:107–117

Koch JC (1917) The laws of bone architecture. Am J Anat 21:177–293

Kuhn JL Goldstein SA, Choi K, London M, Feldkamp LA, Matthews LS (1990) Comparison of the trabecular and cortical tissue moduli from humal iliac crests. J Orthop Res 8:833–842

Kuiper JH, Huiskes R, Weinans H (1992) Bone remodeling: self optimization vs. global optimization. Trans Orthop Res Soc 17:

Kummer B (1962) Funktioneller Bau und funktionelle Anpassung des Knochens. Anat Anz 111:261–293

Kummer B (1963) Grundlagen der Biomechanik des menschlichen Stütz und Bewegungsapparates. Congress of the Society of International Surgical Orthopedic Traumatology, vol II. Vienna, pp D65–D88

Kummer B (1968) Die Beanspruchung des menschlichen Hüftgelenks. I. Allgemeine Problematik. Z Anat Entw Gesch 127:277–285

Kummer B (1971) Computer simulation of the adaptation of bone to mechanical stress. In: Proceedings of the San Diego Biomedical Symposium, San Diego, pp 5–12

Kummer B (1972) Biomechanics of bone: mechanical properties, functional structure, functional adaptation. In: Fung YC (ed) Biomechanics. Prentice-Hall, Englewood Cliffs, pp 237–271

Kummer B (1974) Biomechanik der Gelenke (Diarthrosen). Die Beanspruchung des Gelenkknorpels. Wiss. Konf. Deutscher Naturforscher u. Ärzte. Biopolymere und Biomechanik von Bindegewebssystemen. Springer Berlin Heidelberg New York, pp 19–28

Kummer B (1985) Einführung in die Biomechanik des Hüftgelenks. Springer Berlin Heidelberg New York

Kummer B, Breul R, Stauss M, Lohscheidt K (1987) Spannungsverteilung über Kugelgelenkflächen. Verh Anat Ges 81:445–446

Kurrat HJ, Oberländer W (1978) The thickness of the cartilage in the hip joint. J Anat 126:145–455

Kurrat HJ, Oberländer W (1981) Die Knorpeldickenverteilung am proximalen Anteil des menschlichen Ellenbogengelenkes. Eine funktionelle Analyse. Morphol Med 1:15–24

Lamarck JB (1809) Philosophie zoologique. Bailliere, Paris

Lanyon LE (1984) Functional strain as a determinant for bone remodeling. Calcif Tissue Int 36 [Suppl 1]:S56–S61

Lanyon LE (1993) Osteocytes, strain detection, bone modeling and remodeling. Calcif Tissue Int 53 [Suppl 1]:S102–S106

Lanyon LE, Rubin CT (1984) Static vs dynamic loads as an influence on bone remodelling. J Biomech 17:897–905

Lanyon LE, Hampson WG, Goodship AE, Shah JS (1975) Bone deformation recorded in vivo from strain gauges attached to the human tibial shaft. Acta Orthop Scand 46 (2):256–268

Lelkes G (1958) Experiments in vitro on the role of movement in the development of joints. Embryol Exp Morphol 6:183–186

Lidgren L (1996) Presidential address. Trans Eur Orthop Res Soc 6:IX–X

Löhe F, Eckstein F, Sauer T, Putz R (1996) Structure, strain and function of the transverse acetabular ligament. Acta Anat 157:315–323

Lösch A, Eckstein F, Haubner M, Englmeier KH (1997) A non invasive technique for 3 dimensional assessment of articular cartilage thickness based on MRI. I. development of a computational method. Magn Reson Imaging 15:795–804

Lotz JC, Gerhart TN, Hayes WC (1990) Mechanical properties of trabecular bone from the proximal femur: a quantitative CT study. J Comput Assist Tomogr 14:107–114

Mac Connail MA (1950) The movement of bones and joints. 3. The synovial fluid and its assistants. J Bone Joint Surg Br 32:244–252

Macirowski T, Tepic S, Mann RW (1995) Cartilage stresses in the human hip joint. J Biomech Eng 116:10–18

Marotti G, Cane V, Palazzini S, Palumbo C (1990) Structure function relationship in the osteocyte. Ital J Min Electr Met 4:93–106

Marshall KW, Guthrie BT, Mikulis DJ (1995) Quantitative cartilage imaging. Br J Rheumatol 34 Suppl 1:29–31

Martin RB (1984) Porosity and specific surface of bone. Crit Rev Biomed Eng 10:179–222

Martin RB (1982) The effects of geometric feedback in the development of osteoporosis. J Biomech 5:447–455

Martin RB, Burr DB (1982) A hypothetical mechanism for the stimulation of osteonal remodelling by fatigue damage. J Biomech 15:137–139

Mc Master JH, Weinert CR (1972) Effects of mechanical forces on growing cartilage. Clin Orthop 72:308–314

McNamara BP, Taylor D, Prendergast PJ (1997) Computer prediction of adaptive bone remodelling around noncemented femoral prostheses: the relationship between damage-based and strain-based algorithm. Med Eng Phys 19:454–463

Menton DN, Simmons DJ, Chang SL, Orr BY (1984) From bone lining cell to osteocyte an SEM study. Anat Rec 209:29–39

Merz B, Niederer P, Müller R, Rüegsegger P (1996) Automated finite element analysis of excised human femora based on precision QCT. J Biomech Eng 118:387–390

Meyer von GH (1867) Die Architektur der Spongiosa. Arch Anat Physiol Wiss Med 34:615–628

Milz S, Eckstein F, Putz R (1995) The thickness of the subchondral plate and its correlation with the thickness of the uncalcified articular cartilage in the human patella. Anat Embryol 192:437–444

Milz S, Eckstein F, Putz R (1997) The thickness distribution of the subchondral mineralization zone of the trochlear notch and its correlation with the cartilage thickness: an expression of functional adaptation to mechanical stress acting on the humero-ulnar joint. Anat Rec 248:189–197

Mitrovic D (1982) Development of the articular cavity in paralyzed chick embryos and in chick embryo limb buds cultured on chorioallantoic membranes. Acta Anat (Basel) 113:313–324

Miyanaga Y, Fukubayashi T, Kurosawa H (1984) Contact study of the hip joint. Load deformation pattern, contact area and contact pressure. Arch Orthop Trauma Surg 103:13–17

Mizrahi J, Solomon L, Kaufman B, Duggan TO (1981) An experimental method for investigating load distribution in the cadaveric human hip. J Bone Joint Surg Br 63:610–613

Mizrahi J, Maroudas A, Lanir Y, Ziv I, Webber TJ (1986) The "instantaneous" deformation of cartilage: effects of collagen fiber orientation and osmotic stress. Biorheology 23:311–330

Mockenhaupt J (1990) Pressure distribution in partly contacting joints a computerized simulation model. Anat Anz 171:313–321

Molzberger H (1973) Die Beanspruchung des menschlichen Hüftgelenks. IV. Analyse der funktionellen Struktur der Tangentialfaserschicht des Hüftpfannenknorpels. Z Anat Entwicklungsgesch 139:283–306

Morrey BF (1992) The elbow and its disorders, 2nd edn. Saunders Philadelphia

Mow VC, Ratcliffe A (1997) Structure and function of articular cartilage and meniscus. In: Mow VC, Hayes WC (eds) Basic orthopaedic biomechanics, chap 4, 2nd edn. Lippincott Raven, New York

Mow VC, Kuei SC, Lai WM, Armstrong CG (1980) Biphasic creep and stress relaxation of articular cartilage in compression? Theory and experiments. J Biomech Eng 102:73–84

Mow VC, Holmes MH, Lai WM (1984) Fluid transport and mechanical properties of articular cartilage: a review. J Biomech 17:77–94

Mow VC, Gibbs MC, Lai WM, Zhu WB, Athanasiou KA (1989) Biphasic indentation of articular cartilage II. A numerical algorithm and an experimental study. J Biomech 22:853–861

Mow VC, Ateshian GA, Spilker RL (1993) Biomechanics of diarthrodial joints: a review of twenty years of progress. J Biomech Eng 115:460–467

Muccio J, Tsuzaki M, Yang X, Banes E (1999) Influence of strain rate on gap junction expression in osteoblast-like cells. Trans Orthop Res Soc 23:33

Mühlbauer R, Lukasz S, Faber S, Englmeier K-H, Reiser M, Eckstein F (2000) Knorpelvolumina im Kniegelenk von Sportlern und Nicht-Sportlern – eine quantitative Analyse mittels Magnetresonanztomographie. Sportorthopädie Sporttraumatologie 14:117–121

Mühlbauer R, Lukasz S, Faber S, Stammberger T, Eckstein F (1999) Comparison of knee joint cartilage thickness in triathletes and physically inactive volunteers – 3D analysis with magnetic imaging. Am J Sports Med: (in press)

Müller-Gerbl M, Schulte E, Putz R (1987) The thickness of the calcified layer of articular cartilage: a function of the load supported? J Anat 154:103–111

Müller-Gerbl M, Putz R, Hodapp N, Schulte E, Wimmer B (1989) Computed tomography osteoabsorptiometry for assessing the density distribution of subchondral bone as a measure of long term mechanical adaptation in individual joints. Skeletal Radiol 18:507–512

Müller-Gerbl M, Putz R, Kenn R (1992) Demonstration of subchondral bone density patterns by three dimensional CT osteoabsorptiometry as a noninvasive method for in vivo assessment of individual long term stresses in joints. J Bone Miner Res [Suppl 2]:S411–S418

Müller-Gerbl M, Putz R, Kenn R, Kierse R (1993) People in different age groups show different hip joint morphology. Clin Biomech 8:66–72

Müller-Gerbl M, Putz R (1995) Functional anatomy of the ankle joint. In: Heim FA (ed) The pilon tibial fracture. Saunders, Philadelphia

Müller-Gerbl M (1998) The subchondral bone plate. Adv Anat Embryol Cell Biol 141:1–134

Müller R, Rüegsegger P (1995) Three dimensional finite element modelling of non invasively assessed trabecular bone structures. Med Eng Phys 17:126–133

Müller R, Hildebrand T, Ruegsegger P (1994) Non invasive bone biopsy: a new method to analyse and display the three dimensional structure of trabecular bone. Phys Med Biol 39:145–164

Müller R, Hahn M, Vogel M, Delling G, Ruegsegger P (1996) Morphometric analysis of noninvasively assessed bone biopsies: comparison of high resolution computed tomography and histologic sections. Bone 18:215–220

Müller R, Van Campenhout H, Van Damme B, Van Der Perre G, Dequeker J, Hildebrand T, Ruegsegger P (1998) Morphometric analysis of human bone biopsies: a quantitative structural comparison of histological sections and micro-computed tomography. Bone 23:59–66

Mullender MG, Huiskes R (1995) Proposal for the regulatory mechanism of Wolff's law. J Orthop Res 13:503–512

Mullender MG, Huiskes R, Weinans H (1994) A physiological approach to the simulation of bone remodeling as a self organizational control process. J Biomech 27:1389–1394

Mullender MG, Huiskes R (1997) Osteocytes and bone lining cells: which are the best candidates for mechano sensors in cancellous bone? Bone 20:527–532

Mullender MG, van Rietbergen B, Ruegsegger P, Huiskes R (1998) Effect of mechanical set point of bone cells on mechanical control of trabecular bone architecture. Bone 22:125–131

Münsterer O, Eckstein F, Hahn D, Putz R (1996) Computer aided 3D assessment of human knee cartilage in Vitro and in Vivo. Clin Biomech 11:260–266

Murray PD (1926) An experimental study of the development of the limbs of the chick. Proc Linnean Soc New South Wales 51:187–263

Murray PDF, Selby D (1930) Intrinsic and extrinsic factors in the primary development of the skeleton. Roux Arch Entw Mech 122:629–662

Murray PD, Drachman DB (1969) The role of movement in the development of joints and related structures: the head and neck in the chick embryo. J Embryol Exp Morphol 22:349–371

Neidlinger Wilke C, Stalla I, Claes L, Brand R, Hoellen I, Rubenacker S, Arand M, Kinzl L (1995) Human osteoblasts from younger normal and osteoporotic donors show differences in proliferation and TGF beta release in response to cyclic strain. J Biomech 28:1411–1418

Oberländer W (1973) Die Beanspruchung des menschlichen Hüftgelenks. V. Die Verteilung der Knochendichte im Acetabulum. Z Anat Entwicklungsgesch 140:367–384

Oberländer W, Kurrat HJ (1982) Die Knorpeldickenverteilung im distalen Teil des Ellenbogengelenks und ihre funktionelle Deutung. Morphol Med 2:45–52

Odgaard A (1997) Three-dimensional methods for quantification of cancellous bone architecture. Bone 20:315–328

Oloyede A, Flachsmann R, Broom ND (1992) The dramatic influence of loading velocity on the compressive response of articular cartilage. Connect Tissue Res 27:211–224

Ostergaard K, Salter DM (1998) Immunohistochemistry in the study of normal and osteoarthritic articular cartilage. Progr Histochem Chytochem 33:93–168

Owan I, Burr DB, Turner CH, Qiu J, Tu Y, Onyia JE, Duncan RL (1997) Mechanotransduction in bone: osteoblasts are more responsive to fluid forces than mechanical strain. Am J Physiol 273:C810–C815

Palumbo C, Palazzini S, Marotti G (1990) Morphological study of intercellular junctions during osteocyte differentiation. Bone 11:401–406

Parfitt AM (1982) The coupling of bone formation to bone resorption: a critical analysis of the concept and of its relevance to the pathogenesis of osteoporosis. Metab Bone Dis Relat Res 4:1–6

Parfitt AM (1994) Osteonal and hemi osteonal remodeling: the spatial and temporal framework for signal traffic in adult human bone. J Cell Biochem 55:273–286

Parfitt AM (1983) The physiologic and clinical significance of bone histomorphometric data. In: Recker RR (ed) Bone histomorphometry: techniques and interpretation. CRC, Boca Raton, pp 143–222

Pauwels F (1955) Die Bedeutung der funktionellen Anatomie für die Orthopädie am Beispiel des coxalen Femurendes. Presented at the Congress of the Deutsche Orthopädie Gesellschaft, Hamburg

Pauwels F (1959) Die Struktur der Tangentialfaserschicht des Gelenkknorpels der Schulterpfanne als Beispiel für ein verkörpertes Spannungsfeld. Z Anat Entw Gesch 121:188–240

Pauwels F (1960) Eine neue Theorie über den Einfluß mechanischer Reize auf die Differenzierung der Stützgewebe. Z Anat Entw Gesch 121:478–515

Pauwels F (1963) Die Druckverteilung im Ellbogengelenk, nebst grundsätzlichen Bemerkungen über den Gelenkdruck. Z Anat Entw Gesch 123:643–667

Pauwels F (1965) Gesammelte Abhandlungen zur funktionellen Anatomie des Bewegungsapparates. Springer Berlin Heidelberg New York

Pauwels F (1973) Atlas zur Biomechanik der gesunden und kranken Hüfte. Springer Berlin

Pauwels F (1980) Biomechanics of the locomotor apparatus. Springer, Berlin Heidelberg New York

Pead MJ, Suswillo R, Skerry TM, Vedi S, Lanyon LE (1988) Increased 3H uridine levels in osteocytes following a single short period of dynamic bone loading in vivo. Calcif Tissue Int 43:92–96

Pelegrini O (1933) Lo sviluppo di abbozzi di articolazioni impiantati della membrana corioallantoide. Atti Soc Med Padova 11:927–941

Persson M (1983) The role of movements in the development of sutural and diarthrodial joints tested by long term paralysis of chick embryos. J Anat 137:591–599

Peterfy CG, van Dijke CF, Janzen DL, Gluer CC, Namba R, Majumdar S, Lang P, Genant HK (1994) Quantification of articular cartilage in the knee with pulsed saturation transfer subtraction and fat suppressed MR imaging: optimization and validation. Radiology 192:485–491

Peterfy CG, van Dijke CF, Lu Y, Nguyen A, Connick TJ, Kneeland JB, Tirman PF, Lang P, Dent S, Genant HK (1995) Quantification of the volume of articular cartilage in the metacarpophalangeal joints of the hand: accuracy and precision of three dimensional MR imaging. AJR Am J Roentgenol 165:371–375

Peyron JG (1986) Osteoarthritis. The epidemiologic viewpoint. Clin Orthop 213:13–19

Pidaparti RMV, Turner CH (1997) Cancellous bone architecture: advantages of nonorthogonal trabecular alignment under multidrictional joint loading. J Biomech 30:979–983

Pitsillides AA, Rawlinson SC, Suswillo RF, Bourrin S, Zaman G, Lanyon LE (1995a) Mechanical strain induced NO production by bone cells: a possible role in adaptive bone (re)modeling? FASEB J 9:1614–1622

Pitsillides AA, Archer CW, Prehm P, Bayliss MT, Edwards JC (1995b) Alterations in hyaluronan synthesis during developing joint cavitation. J Histochem Cytochem 43:263–273

Prendergast PJ, Huiskes R (1995) An investigation of Pauwels' mechanism of tissue differentiation. Trans Eur Orthop Res Soc 5:76

Prendergast PJ (1997a) Finite element models in tissue mechanics and orthopaedic implant design. Clin Biomech 12:343-366

Prendergast PJ, Huiskes R, Soballe K (1997b) Research Award 1996. Biophysical stimuli on cells during tissue differentiation at implant interfaces. J Biomech 30:539-548

Raab Cullen DM, Thiede MA, Petersen DN, Kimmel DB, Recker RR (1994) Mechanical loading stimulates rapid changes in periosteal gene expression. Calcif Tissue Int 55:473-488

Radin EL, Ehrlich MG, Chernack R, Abernethy P, Paul IL, Rose RM (1978) Effect of repetitive impulsive loading on the knee joints of rabbits. Clin Orthop 131:288-293

Radin EL, Burr DB, Caterson B, Fyhrie D, Brown TD, Boyd RD (1991) Mechanical determinants of osteoarthrosis. Semin Arthritis Rheum 21 3 Suppl:12-21

Rapperport DJ, Carter DR, Schurman DJ (1985) Contact finite element stress analysis of the hip joint. J Orthop Res 3:435-446

Raux P, Townsend PR, Miegel R, Rose RM, Radin EL (1975) Trabecular architecture of the human patella. J Biomech 8:1-7

Rawlinson SC, Mosley JR, Suswillo RF, Pitsillides AA, Lanyon LE (1995) Calvarial and limb bone cells in organ and monolayer culture do not show the same early responses to dynamic mechanical strain. J Bone Miner Res 10:225-232

Reich KM, Frangos JA (1991) Effect of flow on prostaglandin E2 and inositol trisphosphate levels in osteoblasts. Am J Physiol 261:C428-C432

Reich KM, Gay CV, Frangos JA (1990) Fluid shear stress as a mediator of osteoblast cyclic adenosine monophosphate production. J Cell Physiol 143:100-104

Reilly DT, Burstein AH (1995) The elastic and ultimate properties of compact bone tissue. J Biomech 8:393-405

Rho JY, Hobatho MC, Ashman RB (1995) Relations of mechanical properties to density and CT numbers in human bone. Med Eng Phys 17:347-355

Rice JC, Cowin SC, Bowman JA (1988) On the dependence of the elasticity and strength of cancellous bone on apparent density. J Biomech 21:155-168

Riede UN, Heitz P, Ruedi T (1971) Gelenkmechanische Untersuchungen zum Problem der posttraumatischen Arthrosen im oberen Sprunggelenk. II. Einfluß der Talusform auf die Biomechanik des oberen Sprunggelenkes. Langenbecks Arch Chir 330:174-184

Robertson SS, Dierker LJ, Sorokin Y, Rosen MG (1982) Human fetal movement: spontaneous oscillations near one cycle per minute. Science 218:1327-1330

Rodan GA, Bourret LA, Harvey A, Mensi T (1975) Cyclic AMP and cyclic GMP: mediators of the mechanical effects on bone remodeling. Science 189:467-469

Rodrigues H, Jacobs C, Guedes JM, Bendse MP Global and local material optimization models applied to anisotropic bone adaptation: synthesis in bio-solid mechanics. International Union of Theoretical and Applied Mechanics (in press)

Roelofsen J, Klein Nulend J, Burger EH (1995) Mechanical stimulation by intermittent hydrostatic compression promotes bone specific gene expression in vitro. J Biomech 28:1493-1503

Roesler H (1987) The history of some fundamental concepts in bone biomechanics. J Biomech 20:1025-1034

Roux W (1881) Der Kampf der Teile im Organismus. Engelmann, Leipzig

Roux W (1895) Gesammelte Abhandlungen über Entwicklungsmechanik der Organismen. Engelmann, Leipzig

Roux W (1912) Terminologie der Entwicklungsmechanik der Tiere und Pflanzen. Engelmann, Leipzig

Ruano-Gil D, Nardi Vilardaga J, Tejedo Mateu A (1978) Influence of extrinsic factors on the development of the articular system. Acta Anat (Basel) 101:36-44

Ruano-Gil D, Nardi Vilardaga J, Teixidor Johe A (1985) Embryonal hypermobility and articular development. Acta Anat (Basel) 123:90-92

Rubin CT, Mc Leod KJ (1995) Endogeneous control of bone morphology via frequency specific, low magnitude functional strain. In: Odgaard A, Weinans H (eds) Bone structure and remodeling. Recent advances in human biology, vol 2. World Scientific, Singapore, pp 79-90

Rubin CT, Lanyon LE (1984) Regulation of bone formation by applied dynamic loads. J Bone Joint Surg Am 66:397-402

Rüegsegger P, Koller B, Muller R (1996) A microtomographic system for the nondestructive evaluation of bone architecture. Calcif Tissue Int 58:24-29

Rushfeldt PD, Mann RW (1979) Influence of cartilage geometry on the pressure distribution in the human hip joint. Science 204:413–415

Rushfeldt PD, Mann RW, Harris WH (1981) Improved techniques for measuring in vitro the geometry and pressure distribution in the human acetabulum. II Instrumented endoprosthesis measurement of articular surface pressure distribution. J Biomech 14:315–323

Sah RL, Kim YJ, Doong JY, Grodzinsky AJ, Plaas AH, Sandy JD (1989) Biosynthetic response of cartilage explants to dynamic compression. J Orthop Res 7:619–636

Salter DM, Robb JE, Wright MO (1997) Electrophysiological responses of human bone cells to mechanical stimulation: evidence for specific integrin function in mechanotransduction. J Bone Miner Res 12:1133–1141

Saxena RK, Sahay KB, Guha SK (1991) Morphological changes in the bovine articular cartilage subjected to moderate and high loadings. Acta Anat Basel 142:152–157

Scale WTD, Kurth A (1993) Computergestützte Analyse der Kinematik des oberen Sprunggelenks. Z Orthop 131:14–17

Schaller MD, Parsons JD (1994) Focal adhesion kinase and associated proteins. Curr Opin Cell Biol 6:705–710

Schenck RC Jr, Athanasiou KA, Constantinides G, Gomez E (1994) A biomechanical analysis of articular cartilage of the human elbow and a potential relationship to osteochondritis dissecans. Clin Orthop 299:305–312

Setton LA, Zhu W, Mow VC (1993) The biphasic poroviscoelastic behavior of articular cartilage: role of the surface zone in governing the compressive behavior J Biomech 26:581–592

Shephard DET, Seedholm BB (1996) Measurement of articular cartilage compressive modulus under physiological loading conditions. In: Proceedings: Symposium on the biology of the synovial joint, Cardiff, p 54

Shiba R, Sorbie C, Siu DW, Bryant JT, Cooke TD, Wevers HW (1988) Geometry of the humeroulnar joint. J Orthop Res 6:897–906

Simkin PA, Graney DO, Fiechtner JJ (1980) Roman arches, human joints, and disease: differences between convex and concave sides of joints. Arthritis Rheum 23:1308–1311

Simkin PA, Heston TF, Downey DJ, Benedict RS, Choi HS (1991) Subchondral architecture in bones of the canine shoulder. J Anat 175:213–227

Simon SR, Radin EL, Paul IL, Rose RM (1972) The response of joints to impact loading. II. In vivo behavior of subchondral bone. J Biomech 5:267–272

Singh I (1978) The architecture of cancellous bone. J Anat 127:305–310

Sittek H, Eckstein F, Gavazzeni A, Milz S, Kiefer B, Schulte E, Reiser M (1996) Assessment of normal patellar cartilage volume and thickness with MRI an analysis of currently available sequences. Skeletal Radiol 25:55–62

Skerry TM, Bitensky L, Chayen J, Lanyon LE 1989) Early strain related changes in enzyme activity in osteocytes following bone loading in vivo. J Bone Miner Res 4:783–788

Smalt R, Mitchell FT, Howard RL, Chambers TJ (1997) Induction of NO and prostaglandin E2 in osteoblasts by wall-shear stress but not mechanical strain. Am J Physiol 273:E751–E758

Smith RL, Thomas KD, Schurman DJ, Carter DR, Wong M, van der Meulen MC (1992) Rabbit knee immobilization: bone remodeling precedes cartilage degradation. J Orthop Res 10:88–95

Soballe K (1993) Hydroxyapatite ceramic coating for bone implant fixation. Mechanical and histological studies in dogs. Acta Orthop Scand [Suppl] 255:1–58

Soltz MA, Ateshian GA (1998) Experimental verification of cartilage fluid pressurization in confined compression creep and stress relaxation. Trans Orthop Res Soc 23:224

Somjen D, Binderman I, Berger E, Harell A (1980) Bone remodelling induced by physical stress is prostaglandin E2 mediated. Biochim Biophys Acta 627:91–100

Soslowsky LJ, Flatow EL, Bigliani LU, Mow VC (1992a) Articular geometry of the glenohumeral joint. Clin Orthop 285:181–190

Soslowsky LJ, Flatow EL, Bigliani LU, Pawluk RJ, Ateshian GA, Mow VC (1992b) Quantitation of in situ contact areas at the glenohumeral joint: a biomechanical study. J Orthop Res 10:524–534

Springer V, Graichen H, Stammberger T, Englmeier K-H, Reiser M, Eckstein F (1998) Nichtinvasive Analyse des Knorpelvolumens und der Knorpeldicke im menschlichen Ellbogengelenk mittels MRT. Ann Anat 180:331–338

Stammberger T, Eckstein F, Englmeier K-H, Reiser M (1999) Determination of 3D cartilage thickness data from MR imaging - computational method and reproducibility in the living. Magn Reson Med 41:529–536

Stanford CM, Stevens JW, Brand RA (1995) Cellular deformation reversibly depresses RT PCR detectable levels of bone related mRNA. J Biomech 28:1419–1427

Stevens SS, Beaupre GS, Carter DR (1997) A mechanobiological model for bone growth and articular cartilage development. Trans Orthop Res Soc 22:522

Stevens SS, Beaupre GS, Carter DR (1999) Computer model of enchondral growth and ossification in long Gones: Biological and mechanobiological influences. I Orthop Res 17: 646-653

Stormont TJ, An KN, Morrey BF, Chao EY (1985) Elbow joint contact study: comparison of techniques. J Biomech 18:329–336

Sullivan GE (1966) Prolonged paralysis of the chick embryo with special reference to effects on the vertebral column. Aust J Zool 14:1–17

Sumner DR, Turner TM, Urban RM, Galante JO (1992) Remodeling and ingrowth of bone at two years in a canine cementless total hip arthroplasty model. J Bone Joint Surg Am 74:239–250

Thompson D'AW (1942) On growth and form, 2nd edn. Cambridge University, London

Tillmann B (1971) Die funktionelle Beanspruchung und Morphologie des menschlichen Ellenbogengelenks. I. Funktionelle Morphologie der Gelenkflächen. Z Anat Entw Gesch 134:328–342

Tillmann B (1971) A contribution to the functional morphology of articular surfaces. Norm Pathol Anat (Stuttgart) 34:1–50

Toma CD, Ashkar S, Gray ML, Schaffer JL, Gerstenfeld LC (1997) Signal transduction of mechanical stimuli is dependent on microfilament integrity: identification of osteopotin as a mechanically induced gene in osteoblasts. J Bone Miner Res 12:1626–1636

Tomatsu T, Imai N, Takeuchi N, Takahashi K, Kimura N (1992) Experimentally produced fractures of articular cartilage and bone. The effects of shear forces on the pig knee. J Bone Joint Surg Br 74:457–462

Turner CH, Forwood MR, Otter MW (1994) Mechanotransduction in bone: do bone cells act as sensors of fluid flow? FASEB J 8:875–878

Turner CH, Owan I, Takano Y (1995) Mechanotransduction in bone: role of strain rate. Am J Physiol 269:E438–E442

Turner TM, Sumner DR, Urban RM, Rivero DP, Galante JO (1996) A comparative study of porous coatings in a weight bearing total hip arthroplasty model. J Bone Joint Surg Am 1986 68:1396–1409

Ulrich D, van Rietbergen B, Laib A, Rueegsegger P (1999) Load transfer analysis of the distal radius from in-vivo high-resolution CT-imaging. J Biomech (in press)

Urban JP (1994) The chondrocyte: a cell under pressure. Br J Rheumatol 33:901–908

Van Buskirk WC, Ashmann RB (1981) The elastic moduli of bone. In: Cowin SC (ed) Mechanical properties of bone. American Society of Medical Engineers, New York, 131–143

Vandenburgh HH (1992) Mechanical forces and their second messengers in stimulating cell growth. Am J Physiol 262:350–355

Van der Meulen MC, Beaupre GS, Carter DR (1993) Mechanobiologic influences in long bone cross sectional growth. Bone 14:635–642

Van der Meulen MCH, Beaupre GS, Morey Holton ER, Carter DR (1995) Modeling diaphyseal changes during growth and adaptation. In: Odgaard A, Weinans H (eds) Bone structure and remodeling. Recent advances in human biology, vol 2. World Scientific, Singapore, pp 239–250

Van Rietbergen B, Huiskes R, Weinans H, Sumner DR, Turner TM, Galante JO (1993) ESB Research Award 1992. The mechanism of bone remodeling and resorption around press fitted THA stems. J Biomech 26:369–382

Van Rietbergen B, Weinans H, Huiskes R, Odgaard A (1995) A new method to determine trabecular bone elastic properties and loading using micromechanical finite element models. J Biomech 28:69–81

Van Rietbergen B, Odgaard A, Kabel J, Huiskes R (1996a) The inherent mechanical quality of trabecular bone architecture can be accurately predicted from fabric and apparent density. Trans Orthop Res Soc 21:82

Van Rietbergen B, Mullender MG, Huiskes R (1996b) A three-dimensional model for osteocyte regulated remodeling simulation at the tissue level. In: Middleton J, Jones ML, Pande GN (eds) Computer methods in biomechanics & biomedical engineering. Gordon and Breach, UK

Van Rietbergen B, Müller R, Ulrich D, Ruegsegger P, Huiskes R (1999) Tissue stresses and strain in trabeculae of a canine proximal femur can be quantified from computer reconstructions. J Biomech (in press)

Vasu R, Carter DR, Harris WH (1982) Stress distributions in the acetabular region I. Before and after total joint replacement. J Biomech 15:155-164

Verdonschot N, Huiskes R (1990) FEM analyses of hip prostheses: validity of the 2 D side plate model and the effects of torsion, Transactions of the 7th Meeting of the European Society of Biomechanics, pp 20A

Vigliani F (1955a) Accrescimento e rinnovamento strutturale della compotta in ossa sottratte alle sollecitazoni meccaniche, Nota I. Ricerche sperimentali nel cane. Z Zellforsch 42:59

Vigliani F (1955b) Accrescimento e rinnovamento strutturale della compotta in ossa sottratte alle sollecitazoni meccaniche, Nota II. Ricerche sperimentali nel cane. Z Zellforsch 43:17

Vogt S, Eckstein F, Schön M, Putz R (1999) Vorzugsrichtung der Kollagenfasern im subchondralen Knochen des Hüft- und Schultergelenks. Ann Anat 181:181-189

Volkmann R (1862) Chirurgische Erfahrungen über Knochenbiegungen und Knochenwachstum. Arch Path Anat 24:512

Walmsley T (1928) Articular mechanics of the diarthroses. J Bone Joint Surg Br 10:40-45

Wang H, Ateshian GA (1997) The normal stress effect and equilibrium friction coefficient of articular cartilage under steady frictional shear. J Biomech 30:771-776

Wang N, Butler J, Ingber DE (1993) Mechanotransduction across the cell surface and through the cytoskeleton. Science 260:1124-1127

Ward AC, Pitsillides AA (1998) Developmental immobilization induces failure of joint cavity formation by a progress involving selective changes in glycosaminoglycans synthesis. Trans Orthop Res Soc 23:199

Ward AC, Pitsillides AA (1999) Mechanodependent joint line selective activation of ERK-1 during embryonic limb development. Trans Orthop Res Soc 24:343

Ward AC, Dowthwaite GP, Pitsillides AA (1999) Hyaluronan in joint development. Biochem Soc Trans 27:128-135

Watson SJ, Bekoff A (1990) A kinematic analysis of hindlimb motility in 9 and 10 day old chick embryos. J Neurobiol 21:651-660

Webb CMB, Zaman G, Mosley JR, Tucker RP, Lanyon LE, Mackie EJ (1997) Expression of tenascin-C in bones responding to mechanical load. Bone Miner Res 12:52-58

Weinans H, Huiskes R, Grootenboer HJ (1989) Convergence and uniqueness of adaptive bone remodeling. Trans Orthop Res Soc 14:310

Weinans H, Huiskes R, Grootenboer HJ (1992) The behavior of adaptive bone remodeling simulation models. J Biomech 25:1425-1441

Weinans H, Huiskes R, Van Rietbergen B, Sumner DR, Turner TM, Galante JO (1993) Adaptive bone remodeling around bonded noncemented total hip arthroplasty: a comparison between animal experiments and computer simulation. J Orthop Res 11:500-513

Weinbaum S, Cowin SC, Zeng Y (1994) A model for the excitation of osteocytes by mechanical loading induced bone fluid shear stresses. J Biomech 27:339-360

Werner (1897) Die Dicke der menschlichen Gelenkknorpel. Dissertation, Berlin

Whalen RT, Carter DR, Steele CR (1988) Influence of physical activity on the regulation of bone density. J Biomech 21:825-837

Witte H, Eckstein F, Recknagel S (1997) A calculation of the forces acting on the fossa acetabuli during walking - based on in vivo measurements, kinematic analysis and morphometry. Acta Anat 160:269-280

Wolff J (1892) Das Gesetz der Transformation der Knochen. Hirschwald, Berlin (reprinted by Schattauer, Stuttgart, 1991)

Wong M, Carter DR (1990) A theoretical model of endochondral ossification and bone architectural construction in long bone ontogeny. Anat Embryol Berl 181:523-532

Wynarsky GT, Greenwald AS (1983) Mathematical model of the human ankle joint. J Biomech 16:241-251

Xia SL, Ferrier J (1992) Propagation of a calcium pulse between osteoblastic cells. Biochem Biophys Res Commun 186:1212–1219

Yelin E, Callaghan LF (1995) The economic cost and social and psychological impact of musculoskeletal conditions. Arthritis Rheum 38:1351–1362

Yelin E (1998) The economics of osteoarthritis. In: Brandt K, Doherty M, Lohmander LS (eds) Osteoarthritis Oxford Medical, Oxford, pp 23–30

Zeng Y, Cowin SC, Weinbaum S (1994) A fiber matrix model for fluid flow and streaming potentials in the canaliculi of an osteon. Ann Biomed Eng 22:280–292

Zimmerman NB, Smith DG, Pottenger LA, Cooperman DR Mechanical disruption of human patellar cartilage by repetitive loading in vitro. Clin Orthop 229:302–307

Acknowledgements

We thank Sabine Hillebrand and Mathias Bergmann (Musculoskeletal Research Group, Anatomische Anstalt, Munich) for their help with the biomechanical experiment, Martin Steinlechner (Institut für Anatomie, Innsbruck) and Nikolaus Holzknecht (Institut für Radiologische Diagnostik, Klinikum Großhadern, Munich) for their support in acquiring the CT data, and Marion Schön and Sandra Vogt (Musculoskeletal Research Group, Anatomische Anstalt, Munich) for the assessment of the subchondral split line patterns. Our grateful thanks are also due to Francis Steel for his help in translating the manuscript, Sabine Mühlsimer and Horst Ruß (Anatomische Anstalt Munich) for drawing some of the figures and obtaining the photographs, and Sandra Vogt for secretarial assistance. These studies were supported by a grant from the Deutsche Forschungsgemeinschaft.